D0099009

THE CUSTOM-MADE BRAIN

THE
CUSTOM-MADE BRAIN

Cerebral Plasticity, Regeneration, and Enhancement

Jean-Didier Vincent and Pierre-Marie Lledo

Translated by Laurence Garey

COLUMBIA UNIVERSITY PRESS   *NEW YORK*

Columbia University Press
Publishers Since 1893
New York   Chichester, West Sussex
cup.columbia.edu
Originally published as *Le cerveau sur measure*, copyright © 2012 Odile Jacob
This English-language edition copyright © 2014 Columbia University Press
All rights reserved

Library of Congress Cataloging-in-Publication Data
Vincent, Jean-Didier, author.
[Cerveau sur mesure. English]
The custom-made brain : cerebral plasticity, regeneration, and enhancement /
Jean-Didier Vincent and Pierre-Marie Lledo. — English-language edition.
pages cm
Includes bibliographical references and index.
ISBN 978-0-231-16450-4 (cloth : alk. paper) — ISBN 978-0-231-53421-5 (e-book)
1. Brain—Physiology. 2. Brain stimulation. 3. Neurophysiology.
I. Lledo, Pierre-Marie, author. II. Title.
QP376.V52713 2014
612.8'2—dc23
2013040696

Columbia University Press books are printed on permanent and durable acid-free paper.
This book is printed on paper with recycled content.
Printed in the United States of America

c 10 9 8 7 6 5 4 3 2 1

COVER DESIGN AND ART: Alex Camlin

References to websites (URLs) were accurate at the time of writing. Neither the author
nor Columbia University Press is responsible for URLs that may have expired or
changed since the manuscript was prepared.

In memory of my father who inspired the title of this book.

To my dearest mother for her unfailing love. You alone incarnate

all the injustice of a neurodegenerative disease

become sadly too common.

P-ML

To my children and grandchildren,

whose brains bear the hopes of my genome.

J-DV

CONTENTS

CONTENTS

ACKNOWLEDGMENTS

WE WISH TO THANK all those who, directly or indirectly, have contributed to making this book possible. Without rigid classification and in no particular order, we rely on our memory to recall them with our greatest affection.

First of all we thank all the members of the Perception and Memory Laboratory at the Institut Pasteur for their assistance and their readiness to share the joys of research.

We are grateful to Gilles Gheusi, Pierre-Jean Arduin, Dominique Martinez, and Serge Picaud for having agreed to read the manuscript in whole or in part. Their remarks, on form or fact, have certainly contributed to improving it significantly.

We express our gratitude to two colleagues, Henri Korn and Jean-Pierre Changeux, for their valuable and sustained exchanges, and their ever enlightened advice.

Finally our thanks would not be complete if I had not had the good fortune to have beside me Pascale, Isabel, and Clément, as well as Daniel, Jean-Bernard, Dominique, Gérard, and Christian, with their generous support.

P-ML

Above all, thanks to Pierre-Marie Lledo and all his team at the Institut Pasteur for our mutual understanding.

To Philippe Vernier, my faithful friend, who welcomed me to his research team on Neurobiology and Development at the Alfred Fessard Institute of Neurobiology (INAF), where he so brilliantly succeeded me.

To Odile Jacob and Bernard Gotlieb for their constant support, and Jean-Luc Fidel, without whom this book would not exist.

To Hélène Hryn for her help in using an instrument that terrifies me, the computer.

To my friend Jean-François Moueix for his discreet presence at my side on the paths of the selva obscura.

To the wanderers in the hills, the Chief and the Colonel, without whom the Bordelais countryside would be less beautiful. To the Lysottes, lord and lady of the manor of Gourgue, for their amiable sovereignty.

Finally, to the Verdiers, Mamie, Alain, and Kéké, and to René Biandon, my epigenetic family.

J-DV

TRANSLATOR'S NOTE

WHEN BRIDGET FLANNERY-McCOY asked me if I would be willing to translate *Le cerveau sur mesure* I was particularly happy to accept. I looked back with pleasure to the time in the late 1970s when Jean-Didier Vincent and I served on the Committee of Administration of the Institute of Neurophysiology and Psychophysiology in Marseille, directed by the late Jacques Paillard. I remember Jean-Didier's subtle rigor and his polite but cutting humor.

I have long admired his works, which deal with an aspect of neuroscience of which he has become a master, what I might call the affective brain. He writes less about the brain's basic sensorimotor tasks and more about love, emotion, passion, and instincts, aspects that make the human brain different from that of other species. Jean-Didier Vincent and Pierre-Marie Lledo now offer us yet a different view of the "custom-made" brain and discuss its long evolutionary history, over both the millennia and over the time of its development in the embryo. They explain the importance of our acquiring a head in which to keep this precious brain; they then take a somewhat novel look at nature and nurture. We follow the brain from the cutting room, where the tailor makes it to measure, through the repair workshop, but the authors go even further and make us wonder what the future might hold in terms of augmenting nature's gift: brains enhanced by drugs and computers.

Translation is not just a matter of substituting words: it is an attempt to convey authors' ideas in a new language, and there are pitfalls all along the way. I have modified some of the layout of the original while trying to retain the feel the authors gave to their work. If I have succeeded, then I am pleased: if I have failed it is my responsibility alone, and I beg the authors' indulgence. I have enjoyed translating this fascinating book, and I hope the readers will enjoy it as much.

PERROY, JULY 2013

THE CUSTOM-MADE BRAIN

I

INTRODUCTION

Man and his brain, no big deal!
—ANONYMOUS

ONE OUGHT REALLY TO SAY, "Man and his brain, what a marvel!"
A man does not realize his brain is there, just as he does not feel a well-
fitting suit: he forgets about it. Our head may feel heavy or painful, but,
paradoxically, the brain, our organ of sensation, is without feeling, a soft
mass painless even to the surgeon's scalpel. It is carved from the same
pattern for an entire species, but it expresses each individual's self, in
other words, his mind.

For long the structure of these 1,500 grams (3.3 lbs.) of soft, pinkish-
yellow matter escaped analysis. Bishop Niels Stensen, or Nicolas Steno
(1638–1686), was a great scientist, anatomist, and geologist, as well as a
theologian.[1] He denounced "the pretension of these people [Descartes
and others] who are so prompt to affirm and give you the history of the
brain and the disposition of its parts with the same assurance as if they
had been present at the composition of this marvelous machine and as if
they had penetrated all the designs of its great architect."[2] Today, after
three centuries of anatomy and sixty years of modern neuroscience,[3]
many veils have been lifted concerning the structure of the brain and its
functions. Thanks to Darwin and his theory of the evolution of species, it
is no longer necessary to think of a design by the great architect. Fans of
the mysterious can be assured, however, that the time has not yet come
when we understand where all the secrets of the human mind are hidden.

Builders of machines to read minds will be disappointed when they find they may be nothing more than Ubu Roi's disembraining machines.[4]

The history of the human brain is intimately related to the odyssey of our species, which appeared about 200,000 years ago with a modern brain that has not evolved since. Who are these people who appeared toward the end of the Quaternary? Whether male or female, they had naked skin but hair on their heads. "He" soon learned to hide his penis and protect it from thorns along his path. "She" exhibited her pendulous breasts and fat buttocks as marks of her femininity. They showed neither shame nor modesty; clothing came later, together with morality and law. They experienced emotions but above all felt compassion, a characteristic born of natural selection and fundamental to the development of their species. These creatures with short canines and no claws could never have survived without mutual aid and the capacity to read one another's minds, to guess their intentions and feelings. They walked in step, and their thoughts passed freely from one brain to another. Their eyes lit up by day and chased solitude away. At night they slept under the stars as if they rose from the flames of their fires.

What every gardener knows is true for all men: they are a product of the earth. So it is natural that their fate be linked to that of the soil that bore them. Six or seven million years ago, East Africa underwent a drought that led to the disappearance of the tropical forest and its replacement by woodland savanna. Moving from branch to branch was difficult, and this favored the appearance of bipedalism among the great apes that began to move at ground level.[5] They were the forebears of humanity with a brain of only 400 grams (just under a pound). The famous Lucy was the jewel of the genus *Australopithecus*, which inhabited the earth at the end of the Tertiary. Later came the genus *Homo*. *Homo habilis* populated Africa, and another species, *Homo erectus*, colonized Europe two million years ago.

At the beginning of this period of prehistory we only find a handful of strange individuals. They stood upright with their heads squarely on their shoulders. They walked straight ahead, their eyes watching the world with amazement.[6] They were dispersed[7] over a territory corresponding to the southern part of modern Ethiopia, in the region of the lower Omo valley near Lake Turkana.[8] These *Homo habilis* were masters

of techniques for fashioning stone and flint into blades and other tools. Two million years before our modern era, these pilgrims had slowly but successfully conquered the earth. This "handy" man was not yet an intellectual: for that he had to wait until his brain doubled in size.[9] Then his priority became to learn how to use his head. To face climatic change and find food far from his land of origin, *Homo* chose the intellectual argument, to develop his brain, but he also adopted a varied diet, which included meat. This omnivorous lifestyle and opportunism were as important as the intellectual approach. In fact, they were linked to it. As an adaptation to carnivorism, the canine teeth of *Homo* shortened, gradually to be replaced by tools and by an increased tolerance to fellow man: he preferred to spend his time perfecting his hunting technique and sharing his food rather than biting others.

With 900 grams (2 lbs.) of brain, *Homo erectus* invented society by sharing work, domesticating fire, cooking food, and fashioning clay. Once again, the nature of the earth and the climate produced changes in human development. There were nine ice ages between 900,000 years ago and 15,000 years ago, and they have left traces even today of the slow climatic oscillations that have marked the surface of our planet.[10] The first men who arrived in Europe, including the famous Cro-Magnons, were trapped by glaciations. These long ice ages had profoundly modified the countryside, the contours of the earth, and sea level; across the world, a land bridge more than a thousand kilometers (six hundred miles) long linked the continents between Alaska and Siberia.

Paleolithic archeology has shed light on these periods, which were so rich in the fundamental events that sealed the destiny of modern man. Paleoanthropologists report that we must have had four common ancestors belonging to the hominins.[11] Their classification has been made possible by the development of anthropological genetics. For instance, we can easily distinguish Neanderthal man from modern *Homo sapiens* by their genes, but do they really represent two separate species?[12] The question remains open, given that certain modern Europeans possess up to 2 percent of Neanderthal genes. This Neanderthal contribution to the modern human genome is small but undeniable and shows that interbreeding was possible between Neanderthals and modern man.[13] Are we in fact *Homo neandersapiens*?

Homo floresiensis was discovered in September 2003 in a cave on the island of Flores in Java, Indonesia. It complicates the issue. This hominin has a small brain (380 cc; 23 cu in), closer in volume to that of a chimpanzee than that of modern man. Yet this primitive man was already capable of making tools as complex as those of his more modern cousin. The mystery remains. Finally, a girl who lived some 50,000 years ago was discovered at the Denisova archeological site in southern Siberia.[14] Is she a new species of hominin or a subspecies of Neanderthal? Speculation about the existence of Denisovans is rife. Some see here the proof of the existence of an Asiatic Neanderthal quite distinct from *Homo*. On the other hand, others see evidence of a new species living before the Neanderthals, a sort of pre-Neanderthal. Once again, genetic analysis provides the surprises. Denisovans apparently mated with *Homo sapiens*, for genetic analysis of modern Melanesians and New Guineans have demonstrated the presence of Denisovan genes. We must accept that human evolution did not follow only one branch but rather that after several dead ends only one led to *Homo sapiens*.

From this complex evolutionary tree we shall concentrate on two species (or varieties?): *Homo sapiens*, modern man, and *Homo neandertalensis*, often imagined as something of a beast. Indeed, the latter had a strange appearance. They were short but had a large head protruding backward; their nose was protuberant, their cheeks flattened; their low forehead ended with a bony ridge overhanging their orbits; their imposing frame was broad and muscular, bearing witness to their strength. These features earned them this reputation for bestiality; on the contrary, their graves and the relics they contain suggest a deep spirituality. Their brain weighed some 1,600 grams (3.5 lbs.), more even than that of modern man, suggesting well-developed intellectual capacities, as is confirmed by their mastery of tool making and the presence of manufactured objects such as jewelry and vessels. Neanderthals disappeared, and we do not know the reason for their extinction. But for the human brain modern times had arrived, and man had to make the most of this organ that evolution had bequeathed. So 35,000 years ago, *Homo sapiens* took over the world and rid it of all other hominin groups, including Neanderthals.

The average size of the adult human brain varies according to individuals and sex but, in spite of recurrent ideological polemics, we must

emphasize that there is no significant correlation between size, ethnic origin, and individual intellectual faculties. On the other hand, this marvel of complexity, with its billions of cells, is in no way immutable or fixed, like the components of a computer. Its matter is built to change and only exists because of change. This means that it incarnates the future. It confers the faculty to accomplish tomorrow operations that we are incapable of realizing today or do things today of which we were incapable yesterday. All our particular skills, manual or intellectual, that conspire to make each of us a unique specialist are, to a great extent, fashioned during the first stages of the brain's development in childhood and adolescence.

The growth of the modern human brain has two important characteristics that we do not find in other mammals, even in other primates. First, the brain needs some two decades to be complete. This slow growth offers the possibility for a long period of education, during which a central feature is instruction. The second is illustrated by the late development of the brain in the newborn. At birth the brain is hardly a quarter of its adult size. So man is born with a double paradox: an immature brain that is in no hurry to make up for lost time.[15] We call this property *secondary altriciality*.[16] During this long growth period the child receives signals from the outside world, interacts with its social group, and can acquire a new function: articulate speech. Nonhuman primates develop according to very different modalities. For example, the volume of the chimpanzee's brain at birth is already some 50 percent of that of the adult, and its growth is over quickly, at around two years of age.

Thanks to the traces left by our environment on our neural circuits, each of us can develop a unique character. Our evolutionary heritage is a nervous system fashioned by the double action of experience and environment. This is the central theme of our story, and we shall discuss it in more detail in chapter 4.

The history of species is of capital importance in understanding how our brain works. The nervous system is contingent on the development of the animal kingdom. We shall see in chapter 2 that the first vertebrates, driven by hunger, perfected predation thanks to adopting a new "head," which emerged from the front of their body and enabled them to move efficiently to seize prey. According to this principle it was locomotion, not sensation, that drove the development of the vertebrate nervous system, which became "central." An efficient perception-action loop linking sensory receptors to muscles was the vital function on which evolution exerted pressure for our brain to develop.

To build a human brain required almost one and a half billion years of evolution. During three-quarters of this time the elaboration of a primitive nervous system enabled animals to acquire a greater degree of sensory and motor autonomy. Until the Jurassic, animals could only move to hunt prey or fight predators.[17] Only much later did cognitive functions like language and symbolic thought appear to enable the immense qualitative leap that paved the way for the emergence of the brain of modern man and its unique qualities for abstraction. Such new mental faculties required a flexible, "plastic" nervous system, not a prewired one. Indeed, acquiring our manual and intellectual skills depends on cerebral machinery in perfect order and properly organized. However, at the same time this organization must be, at least in part, adaptable and reconfigurable at all times and at all ages.

This cerebral plasticity is unquestionably at its most spectacular in the child, but it does not disappear in the adult.[18] We consider that there are two major periods of cerebral adaptability. The first, the *critical period*, corresponds to a window in time during which neural wiring is established so that the brain can acquire the components necessary for its activity and its final structure. At this time, sensory experience is crucial. By demonstrating that the visual cortex develops very early, under the influence of visual experience, the Nobel laureates David Hubel and Torsten Wiesel provided the first neurobiological evidence for the existence of a critical period, even if ethologists like Konrad Lorenz and Nikolaas Tinbergen had invoked it since the 1930s.[19]

During this particular period of brain development, the duration of which depends on the specific function (such as vision, audition, walking, language, or mathematics), the brain is the seat of intense changes that manage this immense construction site. During a very early period, the brain is passive; its circuits are waiting to be fed by sensory or motor stimulation provided by the environment in specific affective contexts. The developmental delay in the human fetus leaves the brain highly vulnerable to conditions in the world in which it is growing.

François Truffaut's film *The Wild Child*, inspired by the story of Victor de l'Aveyron, bears witness to the damage caused when essential experience does not materialize. Based on "The Memorandum and Report on Victor de l'Aveyron" by Jean Itard (1806), this film relates the capture of a deaf and dumb boy, who was running on all fours in the forests of Aveyron in France. Once captured by villagers, this young "savage" was taken to an institution for deaf-mutes in Paris, where he became an object of curiosity for numerous visitors. The neurologist Philippe Pinel considered Victor as an incurable idiot and even tried to intern him in the mental asylum at Bicêtre. The young Dr. Itard, from the deaf-mute institution, saved him from confinement. He persuaded Pinel to entrust him with care of the child, whom he believed capable of education. Unfortunately for him, despite developing surprising mental faculties, Victor never learned to speak. This true story illustrates to what point the human brain, so immature at birth, remains vulnerable to the demands of the environment. In other words, the neotenic brain of the child is very receptive to inscribing the world in its circuits, even decades after birth, as long as the stimulation is adequate.[20]

As it reaches maturity, the young adult's brain becomes more and more refractory to learning from experience. Learning new things is certainly never impossible, but it becomes more difficult. Certain connections remain sufficiently plastic for the phenomenon of learning to influence them throughout a whole lifetime. This second period, adult neuroplasticity,[21] is characterized by improvement of the cerebral machine even after it has acquired a broad repertory of sensory and motor expertise: the postjuvenile brain is not a clean slate upon which newly learned skills are inscribed. This period begins in late childhood and lasts until the individual dies. During this second phase, the brain is no longer passive. It uses strategies to decipher the significance of sensory and motor inputs that stimulate its circuits. In short, it seeks to make sense of a lifetime of experience. The processes of attention, that is to say, the opening of our senses to external or internal reality, are corollary to those of learning. The brain realizes this operation of learning through its ability to turn its attention to evaluating whether the desired aim of specific behavior has been achieved or if the individual has been rewarded by the outcome of a planned behavioral pattern. So it is in this context of attention and

"wanting" that adult neural networks can be reconfigured to maximize the chances that a beneficial situation can happen again and again.

There is no doubt that the rich diversity of human personality, aptitude, and behavior depends in large part on the specific wiring of each individual's brain. Human neurobiological variation is the result of inherited characters but also of learning and environmental influences. As we shall see, the first stages of the building of cerebral circuits remain broadly dependent on genetically programmed cellular and molecular processes. However, once the main lines of the wiring are in place, the precision of nervous activity gradually improves by adding or selectively suppressing, connections in the developing brain. Interaction between the outside world and neural activity provides a mechanism by which the environment can influence both the form and function of the brain to produce a unique individual, emancipated and capable of adapted but unpredictable responses.

Study of the evolution of the modern human brain shows us that individuals have evolved and continue to evolve in terms of increasing and diversifying information exchange with their physical and biological environment. An organism becomes an ever more complete representation of its environment. This vital adaptation to one's environment affects first and foremost the nervous system, for it alone allows integration and management of information from the outside world. In other words, to understand the genesis of an individual as the result of cognitive processes (such as perception, language, memory, or consciousness) we need to understand how history is inscribed in the nervous system. We shall come back to how experience shapes neural circuits and how these changes modify mental faculties. In chapter 3 we shall describe how neural networks are laid down in the embryo by skillfully choreographed cell migrations. These developmental processes are organized, and even organize themselves, under the double influence of genes and environment.

Since the 1980s the development of techniques borrowed from molecular genetics has furnished ample evidence of the determining power of genes on the development and normal function of the brain. Today we can identify, activate, or suppress the activity of a gene in order to understand its influence on brain function. Nevertheless, although it

is undeniable that genes remain determinant for the construction and function of the brain, it is just as certain that an individual's own activity and experiences, and therefore his learning processes, are able to reconfigure the connectivity of his brain and profoundly modify his behavior. So the brain is the product of the dual action of genes and modifications imposed by life experiences. Genes determine the overall pattern and the brain's initial wiring. By allowing adjustments to the brain's precise organization according to experience, adult neuroplasticity ensures that we are able to adapt but also retain our individuality and freedom.

A fundamental principle in neuroscience to which we shall return frequently was hinted at by René Descartes (1596–1650) (who spoke of "tubes" rather than neurons): learning was the result of selection and the reinforcement of connections initially made randomly between neurons on the *tabula rasa* of the brain. This concept was adopted at the turn of the twentieth century by psychologists such as Henri Piéron, who in 1923 suggested that "memory, in fact, is nothing but the reinforcement and facilitation of the passage of the nervous impulse along certain paths." Even if this principle does not provide an explanation of the function of memory, it does, on the other hand, elucidate the processes of learning. A few years later, Donald Hebb envisaged the possibility of understanding much of our behavior from this same principle.[22] What he postulated is still relevant to neurobiologists and depends on two fundamental rules:

1. All percepts are represented physically in the brain by activity in a group of neurons called a cell assembly.

2. Two neurons that are active simultaneously end by becoming associated such that activity in one will facilitate activity in the other: this is *Hebb's principle*.

According to this theory, memory happens at the level of the functional connections between neurons, the primary logic elements, the *synapses*, a term coined by Charles Sherrington.[23] Today, theoretical models of cerebral activity adopt and formalize the proposals of three psychologists: William James (brother of the novelist Henry), Henri Piéron, and Donald Hebb. Modern scientists think that the evolution of neural connections with age and experience obeys the same rules that contribute to individual diversity. The two main forces of evolution, variation and selection, would equally work in the management of the development and function of the nervous system.

This fundamental concept in neuroscience, a sort of neuronal Darwinism, as Gerald Edelman described it, postulates that the young brain contains a large number of connections, with each neuron forming synapses with thousands of others. Later, as the individual advances in age and experience, functional synapses stabilize, and others degenerate. As a result, of all the possible pathways between two neural circuits, the most efficient is selected and then consolidated to be used later. This process, called "epigenesis by selective stabilization of synapses" by Jean-Pierre Changeux,[24] reaches its apogee in early childhood. From the eighteenth week of pregnancy, most of the tens of billions of neurons, of which a large number will die, have already been produced and have reached their ultimate destinations. Under the influence of what the fetus experiences in utero and the baby during its first years of life, many redundant or unusable contacts between cells are eliminated while others are welcomed to stay.

Parallel to this synaptic reorganization there is also a morphological reorganization of neuronal networks caused by the production of new neurons in certain regions of the adult brain. This morphological and functional reorganization illustrates the diversity of mechanisms that enable the brain to acquire new information throughout life. The age of the rigid brain, when neuroscience texts paid but timid tribute to modifications in the adult brain, is therefore over. The late 1970s and early 1980s saw the first experimental evidence of sensory and motor plasticity in the adult brain. At that time, several groups explored the consequences of the interruption of sensory inputs on brain function. Michael Merzenich and Jon Kaas used the restructuring of cortical maps (which we shall discuss later) as evidence of the nervous system's power of adaptation. They showed that if a given cortical area is deprived of sensation it would seek other input from adjacent areas. Since then, modern neuroscience has made plasticity one of the central paradigms of the human brain. The brain is seen as an object capable of modifying the organization of its own circuits as a function of experience. We shall discuss this sometimes highly surprising ability in detail in chapter 4. Among other things, it enables the adult brain to make up for insufficiencies to a limited extent, if one can find the means to stimulate this latent potential. In consequence, the notion of a critical period is challenged today by the discovery of adult cerebral plasticity. Certainly, our neural circuits open up to the outside world progressively just before or just after birth, but they never totally lose this faculty, in spite of what Jean Piaget might have thought.[25]

Although we cannot yet fully explain adult neuroplasticity, we can observe and measure it. It preoccupies theoreticians and clinicians more and more. We see the emergence of a new pharmacopoeia aimed at treating various neurological or psychiatric disorders by targeting the mechanisms underlying adult plasticity. An antidepressant such as fluoxetine (Prozac and its generics), if given soon after a cerebrovascular accident (CVA), can improve functional recovery.[26] Another example is the recent discovery at the Massachusetts Institute of Technology that administering a supplement of magnesium increases the capacity of the adult brain to create new connections and so perform better in tasks involving learning and memory.[27] This should appeal to all lovers of chocolate.

This adaptability of circuits makes the brain "informable," informed continuously by sensory and motor information. It also makes it "deformable" because this same information helps reconfigure its own previously established circuits. One extreme example of cerebral plasticity is the continuous production of new neurons in a healthy brain or one that has suffered permanent damage from trauma. Today, considerable experimental effort is devoted to better control of such extreme plasticity to achieve total or partial functional recovery in patients with neurological disorders. In chapter 4, we shall summarize the present state of knowledge in this field, where today's certainty spells tomorrow's doubts.

So where do we stand today on the promises and claims from scientists in search of research support? Can we really envisage recovery of speech lost after a CVA? What hope do stem cells hold in terms of new therapeutic strategies?[28] What does the future hold for advances in regenerative medicine? We shall tackle all these questions in chapter 5, taking care not to offer false hopes, for the road ahead is long and uncertain.

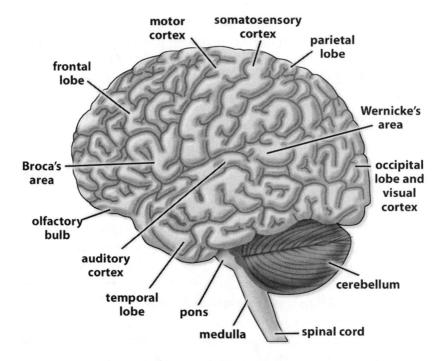

FIGURE 1.1. Lateral view of the human brain. Apart from the cerebellum and the pons and medulla of the brainstem emerging from behind the occipital and temporal lobes, and the beginning of the spinal cord, essentially all we see is cerebral neocortex covering the cerebral hemispheres. Note the various functional areas (motor, somatosensory, visual, auditory, and Broca's and Wernicke's speech areas).

To understand the nature of the processes involved in the reconfiguration of adult neuronal networks it is useful, first, to explain how the different pieces of the puzzle fit together. We shall sketch the broad lines of cerebral architecture in chapter 3 (figure 1.1). In evolutionary terms, human brain function depends on two modules, one *subcortical*, allowing fast but unconscious processing; the second, the *cerebral cortex*, treats information from the internal or external environment consciously but more slowly (figure 1.2). This binary approach, suggested by the English neurologist John Hughlings Jackson (1835–1911), is based on the hierarchical integration of different levels of organization of nerve centers. Jackson's work sought to provide a pathophysiological basis for

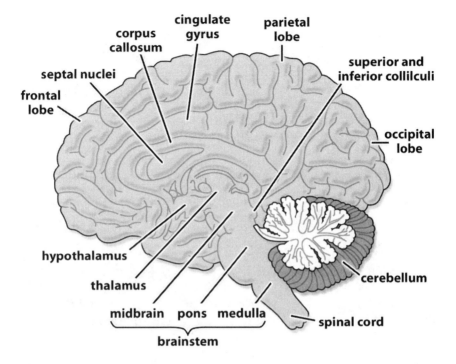

corpus
callosum

cingulate
gyrus

parietal
lobe

septal nuclei

superior and
inferior collilculi

frontal
lobe

occipital
lobe

hypothalamus

thalamus

cerebellum

midbrain   pons   medulla

spinal cord

brainstem

FIGURE 1.2. Medial view of the human brain to show cortical (the lobes and their gyri) and subcortical structures.

neurological and psychiatric disorders. According to anatomical and functional criteria at the time, he proposed a hierarchy of psychic functions and associated pathological conditions with loss of existing functions. This destructuring released underlying negative activity. This concept, formulated in the context of neuropsychiatric disorders, reflected an evolving perspective.

To help understand the essence of this theory, we shall review the phylogenetic aspects of human brain structures in chapter 3. We shall see that the *protoreptilian* brain, the oldest part of the brain, essentially corresponds to the *basal ganglia* (figure 1.3). These neural circuits, which we share with snakes and turtles, participate in autonomic regulation and control the vital functions important for our survival. Thanks to these circuits we can unconsciously alternate between states of waking and sleep, and our respiratory rhythm will adapt, for example, to our

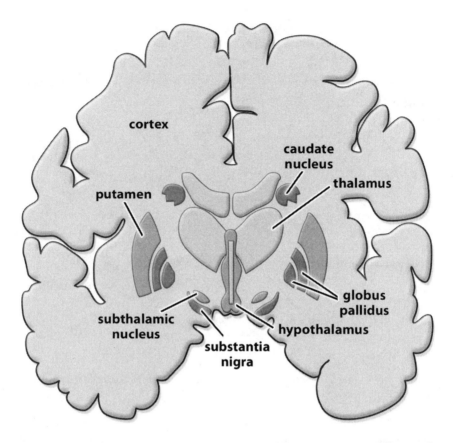

FIGURE 1.3. Diagram of a vertical section through the basal ganglia (caudate nucleus, putamen, globus pallidus, substantia nigra, subthalamic nucleus). It also shows the thalamus and hypothalamus.

climbing stairs without our even being aware of it. This region is in the deepest part of our brain. It is capped by a second *paleomammalian* region, which includes the *limbic system* (figure 1.4) and is shared by all mammals. It is a sort of control tower of our affect that organizes basic instincts such as the expression of emotion and desire. Found even in primitive mammals, it supports the two pillars of the temple of affect: pleasure and suffering. The limbic system is composed of several structures and their related circuits. The first is the *hippocampus* and its related structures, the *mamillary bodies* of the *hypothalamus*, which are important

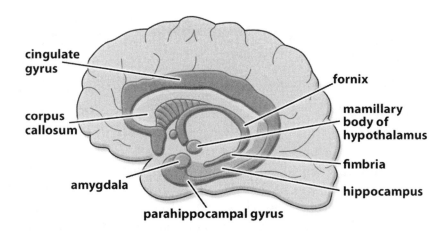

cingulate
gyrus

fornix

corpus
callosum

mamillary
body of
hypothalamus

fimbria

amygdala

hippocampus

parahippocampal gyrus

FIGURE 1.4. Medial view of the human brain to show the limbic system.

for forming memory traces. Next is the *amygdala*. If this nucleus is dysfunctional, the sweetest angel can become a serial killer. Another important limbic structure is the *nucleus accumbens*. This region is responsible for our impulsiveness and controls our motivation. Once activated, these limbic networks release attraction, appetite, and satisfaction. The last subdivision of the paleomammalian brain is the cortex of the *cingulate gyrus* (Broca's *limbic lobe*) and its connections with the thalamus. This structure organizes a wealth of social behavior, such as those expressed in parental relationships.

The limbic brain forms part of the *rhinencephalon* (from the Greek *rhis* for nose and *enkephalos* for brain) of macrosmatic animals, for which olfaction is primary for survival. The emergence of the paleomammalian brain enabled animals to acquire affective reactions, which are useful not only for the protection of the individual but also for the survival of the species. It is the center for emotional aspects of feeding (attractiveness and appetite); for fast action for self-preservation in the face of danger, allowing a decision to fight or to flee; and also for fundamental instincts of sexuality. However, the range of functions stemming from the limbic system would not be complete unless we also considered its participation in mnesic (memory) processes, which bestow an extra degree of freedom on an animal thanks to experience gathered earlier and then retained. The emergence of a powerful memory allowed the development of adaptive conduct of a new type that was not possible through the simple reflex control of the reptilian brain. This possibility of preserving long-term memory traces acquired in a given emotional context explains how each of us can remember precisely what we were doing, for example, during the attacks on the Twin Towers in New York on September 11, 2001.

As we have seen, the limbic system triggers a wide variety of reactions, affective urges, tensions, and basic motivations that are necessary for survival. This deep part of the "ancient" brain facilitates the establishment of comparative and associative links within the mass of incoming information. This comparative function notably supports the development of cortical "association" areas and thus of cognitive abilities. These functions of comparison and association allow situations outside an individual to be handled internally by basic affective reactions.

The third region, the *neocortex* (figure 1.1) of the *neomammalian brain*, envelops the paleomammalian brain. This recent acquisition in evolutionary history culminates in that of man. During this evolution we see a massive increase in the surface area of the neocortex, by a factor of about twenty-five from macaque to man. This region of the brain is the seat of higher cognitive abilities such as language, poetry, and mathematics. Once equipped with a neocortex, the brain was able to acquire conscious sensations such as touch, vision, and audition. New solutions could be found for everyday problems thanks to the circuits of the neocortex and their capacity for constant remodeling.

Certain areas are specialized for such functions as motor activity, touch, vision, or audition, but surrounding them are other, less well-defined areas that integrate a broad spectrum of information, the association areas, which process elementary sensations (perception in its true sense) and identify them (recognition or gnosis). Information from the outside world via the sensory pathways reaches the primary projection areas, but analysis (processing in order to be understood and attain a symbolic value) is undertaken by these cortical association areas. Within the vast heterogeneous territory of the neocortex, the human prefrontal cortex excels by assuming new functions unique to man, such as anticipation and executive planning. With this last evolutionary innovation, complex forms of learning and memory appeared. The neocortex contains ten billion interconnected neurons, which make all these functions possible.

We can add to this classification another form of cortical hierarchy that distinguishes the left and right halves of the brain. This dichotomy is caused by the fact that neural pathways cross the midline, but some do not share information equally to the two sides. Thus, sensation from the left side of the body is received by the right cerebral hemisphere and vice versa. In the same way, motor control of the left side of the body is dependent on the right hemisphere, and vice versa. Language is mainly processed by the left hemisphere, the *dominant hemisphere* in right-handers, and by the right hemisphere in left-handers. In right-handers, the left hemisphere houses the centers for executive motor control, speech, and mathematical reasoning, whereas the right controls more artistic values and intuition. But we must be wary of oversimplification, for we know that both hemispheres constantly exchange information through a transversal highway called the *corpus callosum*.

Two neurobiologists, Roger Sperry and Michael Gazzaniga, demonstrated that if the cerebral hemispheres were separated surgically by transecting the corpus callosum (the *split brain*) to prevent propagation of epileptic fits from one hemisphere to the other, it seemed to have curiously little effect on the patients' behavior. However, these split-brain patients behaved as if they had two minds: each hemisphere could function autonomously. Sperry even suggested that this situation could lead to two different mental states, two states of consciousness each oblivious of the other, a hypothesis that is still hotly debated today. We shall return to this question when we discuss hierarchy in the neocortex in chapter 3.

We have emphasized the extent to which the evolutionary history of the nervous system has increased the mental representation of our body and of the world, which we need to live and to recognize others. So the brain is not an organ like any other: it is a mirror for our perception of ourselves and, more importantly, of others. All our actions and all our sensations are continuously evaluated to measure their individual or collective value. The human brain has developed a unique relationship to others through a system of values. The development of our brain favored not only externalization of our thoughts through language or the use of tools but also our internalization of the thoughts of others. Man is able to experience others' lives within himself and show compassion. As we shall see, the amygdala is a key structure in this context, for it enables us to read others' emotions and to undergo primary conditioning between painful and reinforcing stimuli. It is here that we construct representations and strategies for action as well as form our individual view of the world. Loops of activity between the prefrontal cortex and its underlying subcortical centers enable the brain to attribute moral values to abstract objects or situations. We shall discuss how subcortical structures in rodents that manage punishment and reward have evolved in man toward a more complex system that contributes to our ability to lay down rules of moral conduct, to be moved, to be passionate, or to rebel.

We end on an intentional note of optimism. We wonder about the status of man in years to come. The discovery of cerebral plasticity is at the origin of the new technology now available to the general public: cognitive training and psychostimulants and other "smart drugs" that allow us to optimize brain function. We can enhance it or compensate for certain deficiencies by implanting electrodes to deliver electrical current directly to deep brain structures. Progress in regenerative medicine demonstrates that it is possible to reconstruct many damaged organs, such as skin, blood vessels, and nerves. Techniques initially developed in the theoretical field of nanotechnology are beginning to penetrate the complex domain of the living organism. Although not all brain regions are yet targeted by tissue engineering, it is probable that in the near future these new tools will permit better diagnosis and treatment and, doubtless, improvement of human cognitive function, whether in health or disease. Better control of brain plasticity with cerebral implants enables us to envisage a human being with a superhuman memory, improved brain function, almost perfect night vision, or the ability to control a robot at a distance entirely by thought. This vision of a modifiable adult brain led Ray Kurzweil, the theoretician of transhumanism and technological singularity, to suggest that the body and mind will soon be transcended.[29] According to his calculations, the end of civilization as we know it is thirty years away. Is immortal man already under way?

2

AND THEN THERE WAS SHAPE

To create is likewise to give a shape to one's fate.
—ALBERT CAMUS, *THE MYTH OF SISYPHUS* (1942)

IN SPEAKING OF "CREATING" when considering the origin of the shape of our brain, we risk being associated with the travesty of creationism. But no. The real creator with an effective presence on earth is the human brain: its genius is responsible for tools, from a simple stone or twig to articulate language, which have enabled it to instrumentalize the world and to dispose of it at will. But instead of "origin," we prefer "beginning" in the sense of "In the beginning was the word" or Goethe's "In the beginning was action." Of course, the rationalists' scientific rigor would object to any thought of spiritualism, theology, magic, or the existence of God. The scientist can neither deny nor prove that last concern. When Napoleon was questioning him, the great physicist Pierre-Simon Laplace (1749–1827) did not reply, "God does not exist" but instead, "I had no need of that hypothesis." That is true rigor, but depending on their discipline scientists do not all see it the same way. One of us recently attended a presentation by a mathematician who insisted on the importance of mathematics at high school: it was "magic"! A physicist immediately objected, "In science there is no magic. When I don't know something, I look for rules!" The mathematician, a specialist in absolute rationalism, insisted, "My discoveries are magic. I enchant the world with mathematics, and I enchant myself with theorems!"

So, we refer to the beginning of life on earth nearly four billion years ago without the slightest transcendental context. Life begins when organic molecules recognize one another and unite according to a certain affinity. Life depends on energy extracted from the earth or the sun, which it uses for its own construction. Living matter feeds on living matter. Molecules associate to reveal new forms, testing them, adopting them, or abandoning them under the influence of natural selection. Then the evolution of life begins. The shape of organisms reflects the constancy and stability of species. But there is a strange paradox: their unique identity is accompanied by permanent reshaping, which leads to the disappearance of old species and their replacement by new ones.

As we have it in our heads, as it were, to study the human brain and its antecedents in the evolutionary history of the animal kingdom, we shall first of all tackle the fertile ground of embryology. We shall describe how the brain emerged from a cluster of cells that we call the *morula*.[1] We shall also see how modern genetics, by revealing the existence of a construction plan inscribed within a particular genetic sequence, has surreptitiously reintroduced the ancient concept of shape to the scientific debate. Borrowing from Democritus, for whom shape designated the organization of the parts of a whole,[2] we shall call the anatomical organization of the functional components of the brain *structure*. We invite the reader to tour the principal structures of the human brain or homologous regions of other vertebrates.[3] In discovering the importance of developmental gene expression[4] for the emergence of different shapes and structures in the nervous system, including the encephalon (or brain),[5] modern biology has grasped this question and placed it at the center of contemporary scientific debate.

Geneticists have demonstrated that there exist genes that determine the relative position of organs to one another. These *homeotic* genes organize the precise longitudinal structure of the embryo.[6] Since the work of Thomas Hunt Morgan,[7] fundamental discoveries in genetics have mainly been thanks to experiments on the fruit fly, *Drosophila*, in which homeotic genes were discovered during the 1980s. They work to construct the brain of the embryo and impose its shape, depending on which species we are considering, like a carefully preserved mold. During the evolution of species homeotic genes have thus been responsible for the appearance of distinct anatomical structures along the body axis: a head on top of the thorax, situated itself on top of the abdomen. These anatomical structures are endowed with functional properties that depend as much on their nature as they do on their respective relationships within the organism. As we shall discuss later, invertebrates and vertebrates alike are built according to a general plan that respects these ordered spatial relationships controlled by the action of homeotic genes. Thanks to the conservation of these genes, a chicken's egg will always produce a chick of *Gallus domesticus*, and the human ovum fertilized by a human sperm will always produce a human baby. Modern theories about development teach us that homologous genes that affect the same regions of the genome in organisms as different as flies and men are invariant topological characters, common to all species: hence permanence of form, with variations.

Of course, the idea of the invariance of anatomical structures from one species to another did not wait for the coming of molecular biology to blossom in the minds of scientists. The idea of conservation of form reminds us of the precepts defended by Geoffroy Saint-Hilaire, for whom there existed a single organizational plan for living creatures. From 1796 he formulated the principle of "unity of organic composition," according to which organs, that could vary in size or function from one species to another, occupied a relative position that remained constant from insects to vertebrates. This revolutionary concept was considered as radical at the time, for it opposed the position held by the Academy of Sciences and in particular that of Georges Cuvier.[8]

If there is an aspect of animal life where permanence of shape is most striking in vertebrates, it is in the nervous structures that control desire, passion and affect.[9] As we shall discuss in detail later, the great innovation of the vertebrates in evolution is not so much the vertebra but their "new head,"[10] which assembles the sensory organs (eyes, ears, nose) at the front of the body for seeking out and stalking prey, well protected by a bony mask surrounding a mouth for seizing, killing, and swallowing. Many of the elements of the head are derived from a single embryonic structure, the *neural crest*,[11] composed of pluripotent developing cells that, after migration to the body and head, produce notably the bones and connective tissue of the face, with its voracious mouth, and the autonomic nervous system linking the brain to the vital organs. This organization of the head allows vertebrates to look forward at the world, and it provides them with that capacity of investigating that world we call curiosity. This new capacity relies on their nervous system, and therefore on the function of developmental genes, working irregularly over time and privileging certain parts of their body. This results in an extraordinary ability of the organism to relate to its environment. Contingency enters into the construction of the individual, accompanied by increased desire, which becomes the motor for the individual's opening to the world: active subjectivity. In other words, the inquisitive animal looks into the world and, one is tempted to say, the future. Of course, there is no question of finality or intention.

This clear destiny of the vertebrate brain leads us to the human brain. The time has come to open the door to the workshop of the master tailor who creates this custom-made brain.

Behind Diversity in the Animal Kingdom, a Single Plan

How can one not be dazzled by the fabulous diversity of animal species? But underlying it is a basic plan, a pattern like those used by dressmakers and tailors before they embark on variations. This pattern is stored in the genetic heritage of all individuals. The entire set of genes that we call the *genome* is distributed over our chromosomes in the nucleus of the cell, which is, as we have known for a couple of centuries, the basic unit of living creatures. Certain genes function principally to inform cells where to migrate during embryogenesis and define their final position as they form organs (*organogenesis*) in three dimensions: a dorsoventral, an anteroposterior, and a lateromedial (or left-right) axis. For example, in *Drosophila* homeotic genes of the *Hom-C* family have a single function: to ensure specialization along the anteroposterior axis of the different segments of its body. If a gene of this family mutates, the fly may have not one but two heads. In vertebrates, two groups of genes participate in the laying down of two major axes of the embryo: the genes *Hox* and *Pax*[12] define the respective positions of cells along the anteroposterior and dorsoventral axes of the central nervous system.[13]

The relatively old idea of a single organizational plan is today at the origin of a central concept in embryology, that of *zootype*. It depends on particular patterns of expression of genes that carry precise position information at a particular stage of embryonic development of a *taxon*.[14] Now, developmental biology tells us that all *metazoans*[15] have a common developmental period, during which the embryos of various phyla look similar. For example, we all pass through an embryonic stage during which we grow gills like our aquatic predecessors. Equally, like fish embryos, we fleetingly develop temporary *branchial arches*, which in the human embryo form six tiny fleshy sacks slung on each side of the neck, each containing a cartilaginous primordium. It is the same for the *notochord*, a transitory structure that we grow and then break up again to reutilize its building blocks to make intervertebral disks, just as did the ancient builders of churches and castles. A particular stage of development when all embryos look similar is called a *phylotypic* stage.[16] The existence of this inescapable stage depends on the presence of a conserved complex of homeotic genes[17] that establish an organizational plan for all metazoans. They are *Hox* genes, which are probably among the most studied factors in trying to understand interspecies relationships.

When these genes undergo mutations, "homeotic transformations" arise in which one organ is substituted for another. Probably the most detailed reports in the scientific literature concern the transformation of an antenna of a drosophila into a leg.[18] It is this property that makes us think that this family of genes safeguards the identity of an organism.[19] The discovery of the existence of a complex of homeotic genes common to insects and mammals had the effect of an earthquake in the biological community, which had assumed, until the 1980s, that these animals had different organizational plans. For example, we know that the central nervous system of insects, subdivided into three parts, *protocerebrum*, *deutocerebrum*, and *tritocerebrum*, is placed ventrally, whereas that of vertebrates is placed dorsally.[20]

Homeotic genes constitute a very varied group found in all *eukaryotes* (cells with nuclei), that is to say in animals, plants, fungi, and protists (single-cell eukaryotes). Within this vast array of life only *Hox* genes related to *Drosophila* homeotic genes, and probably stemming from the same ancestral gene,[21] form a particular family only present in multicellular animals.

Genetic analysis clearly shows that homeotic genes are grouped in a complex carried on the same chromosome in *Drosophila*. Closer observation shows that their arrangement on this chromosome is not random. On the contrary, these genes are arranged according to a precise spatial order: the position of the *Hox* genes on the chromosome depends on the part of the body they control. This correspondence between the position of different parts of the body and that of *Hox* genes associated with these target regions is called *spatial colinearity*. Genes situated at one end of the DNA strand (labial) act in the region of the head; those at the other end (abdominal) only deal with more posterior parts of the embryo. The parallel between the gene map and the anteroposterior polarity of the body plan was discovered more than thirty years ago by Edward Lewis.[22] This rule applies to all vertebrates and primitive fish, including that great favorite of biologists, amphioxus.[23] All these species belong to the chordate phylum,[24] which separated from the arthropods more than 500 million years ago.[25] The universal nature of the genetic system that regulates embryonic development means that similar genetic functions can be shared among living organisms. This convergence emphasizes, if such were necessary, the probable single origin of the *scala naturae*.

Molecular analysis of *Hox* genes reveals a particularity that biologists did not expect: certain genes possess an almost identical region in common. This repeated sequence, or *homeobox*, has been described in all the homeotic genes of *Drosophila*.[26] Thanks to progress in molecular biology, more than thirty years ago the notion of homology has been extended to structure, first of molecules, then of genes. The search for molecular or genetic substrates presenting characteristics of homology, that is having common sequences and derived from the evolution of the same molecule or gene in a common ancestor, has provided further criteria for homology. These criteria, or traits, have enabled zoologists to establish new relationships between different structures, for example, the relationships between *somites* (the segmental building blocks of the embryonic body) and the central nervous system. These molecular criteria have facilitated the resolution of problems once considered insoluble, such as evidence for relationships between the nervous systems of animals in which it is situated ventrally and those in which it is dorsal, or the development of the vertebrate brain from sensory and cerebral vesicles of protochordates.

For the first time, thanks to the discovery of homeotic genes, relationships between very different structures have suddenly become obvious, like the primordial mouth of agnathans (such as lampreys) and the aortic arches of birds and mammals.

To appreciate the importance of this revolution in biology we might cite work that has enabled us to understand the evolution of the structure of the *Hox* complex in metazoans. The ancestral *Hox* complex of bilateria (metozoans with bilateral symmetry that have three embryonic germ layers)[27] consists of eight to ten genes capable of conferring a precise identity to different regions of the organism along its anteroposterior axis.[28] In a way, the zootype is simply a batch of genetic information that divides the embryo into differentiated territories progressing from front to back, like wagons coupled to a locomotive. Denis Duboule extended this central concept of embryology by suggesting that the control of expression of homeotic genes can follow another rule, that of *temporal colinearity*, enabling the organism to be constructed from front to back.[29] According to this concept, the anterior part of the body would be the first to form; posterior parts would be added progressively one after the other. If this temporal colinearity is not respected, such as after a mutation, the animal's body would contain organs substituted for others. This is called *homeosis*. The German geneticist Richard Goldschmidt called these mutants "hopeful monsters," for they were prone to reveal themselves better adapted than the current norm for the species.[30] In such cases, the altered organ was intact and could be used for new functions in the mutant. Thanks to their spectacular nature, two examples have remained famous in *Drosophila*: the *Antennapedia* mutant, in which legs replace antennae, and *bithorax*, in which the third thoracic segment is transformed into the second segment. In the latter case, the mutant fly has two pairs of wings instead of one. By beginning its construction anteriorly, the embryo reminds us of the evolutionary history (phylogenesis) of vertebrates as they develop their new heads. We shall come back to this important point to understand the construction of a very special organ, the brain, in which man's psychic faculties are housed.

We should recall that vertebrates are chordates, of which the oldest representatives are the *cephalochordates*, dating from the Cambrian period at the beginning of the Paleozoic era.[31] Their body consists of a repetitive series of segments in which it is impossible to distinguish a neck, thorax, abdomen, pelvis, or tail, as in terrestrial vertebrates. The functions of the different regions of the body can only be spatially coordinated by means of an effective communication system, the nervous system. The identity of the different body territories depends therefore on the nature of the neural circuits that innervate them. To satisfy this challenge and enable each body region to be innervated by a particular neural circuit, their construction must follow harmonious rules. This challenge is not easy if we bear in mind the number of animals possessing a true nervous system. How can we build in an invariable and concerted fashion the bodies of more than a million species with a central nervous system?

In spite of the great variety of shapes and the diversity of behavior that neural circuits are supposed to control, anatomical and functional convergence reveals the fundamental building blocks of living creatures. This convergence of structure between very different species implies that *organic* and *sensory* peripheral units be put in place in coordination with nerve centers. This necessity to operate in a concerted way to form a central neural axis (or *neuraxis*) and its peripheral elements is at the origin of *neuronal zootype*,[32] an extension of the concept of zootype that we defined earlier. We shall return shortly to this important notion that allows us to explain the many constraints to respect when we build a nervous system.

This glimpse of developmental biology emphasizes the key role of *Hox* genes in the emergence of developmental and morphological innovations in metazoans.[33] Apart from the rather audacious, for the time, speculations of Geoffroy Saint-Hilaire in 1822, the definition of an animal has long been limited to a functional or behavioral context. A fish swims, a bird flies. The animal was defined as a living creature that could move, feed, and respond to the solicitations of its environment. The discovery of the key role of *Hox* genes enabled us to understand how an organism could be influenced by the expression of genes that controlled the formation of its organs (homeotic genes) rather than by genes that regulated physiological functions. This modern concept, in which form prevails over function, is epistemologically completely opposed to former hypotheses, such as promulgated by Cuvier, that were based on a classification of organisms according to functional criteria. The very complex structure of our brain depends in the end merely on a handful of homeotic genes that are able to orchestrate its development and give it the characteristic form that we see in the adult.

The discovery of the function of homeotic genes in the establishment of shape owes much to the observation of "monstrous" mutants, but it was also fortuitously enriched by the contributions of theoreticians who, like philosophers, tackled the old question of the genesis of shape. So, in the last few decades we have seen the emergence of mathematical theories that formalize the dynamics of the production of shape in the living world. They go under different names, such as *catastrophe theory*, formulated by René Thom, the *dissipative structures* dear to Ilya Prigogine, the *fractals* of Benoit Mandelbrot, and *chaos theory* and its *strange attractors*, formulated by David Ruelle.[34] These considerations brought order and sense to the enormous mass of experimental data that biologists had accumulated in a dispersed way, within which comparisons were difficult because of the variety of experimental models used. It is remarkable to realize that these theoretical models simply stress the conservation of the most important topological characters of the organizational plan common to all animal species. This conclusion supports the theories formulated earlier by D'Arcy Wentworth Thompson (1860–1948) in *On Growth and Form*, which marked a radical turning point in the study of shape in the living body (morphogenesis) based on methods derived from mathematics and physics.[35] What a fine homage by his successors to this brilliant scientist!

When the Zootype Becomes Neuronal

To reconcile the modern definition of the animal kingdom and that based earlier on behavioral characteristics of different species, two French scientists, Jean Deutsch and Hervé Le Guyader, advanced the hypothesis that the concept of zootype could be extended to the formation of the nervous system. The stakes are simple but crucial. Organizational plans for the formation of the body would also be used for the development of the nervous system. It would ensure better coordination of different embryonic systems if nerves were produced according to the same principles as the muscles they innervate. The primary function of zootype genes in bilaterally symmetrical animals would thus be to define precise neuronal pathways in harmony with the different regions of the embryo: a neuronal zootype.[36] To ensure this function, the zootype genes must be capable not only of controlling the spatial distribution of the embryo's cells but also of guiding nerve cell terminals to their targets. The only means by which our intellectual faculties can develop is by such precise orchestration of cell movements in the embryo.

To illustrate the rules of controlled cell migration we shall now study the way in which, starting with an egg, such complex and different organisms as a fly and a human being can develop. Each organism has its genetic program, which controls its development, determines its shape, and finally manages its functions.

The Shapes We Have Inherited

The appearance of neurons is one of the major developments in the evolution of multicellular animals. It occurred at the *eumetazoan* branch, with the sponges remaining the only metazoans without a nervous system.[37] To understand how we inherited a nervous system with such complex yet precise shapes and structures, we must first see how positional information participates in the differentiation of cells.[38] We must recall that the story of an individual begins with the fertilization by a sperm of an oocyte, a tiny cell about a tenth of a millimeter in diameter in man. Fertilization produces an ovum by two major processes. The fusion of the nuclei of these two gametes allows on the one hand the fusion of the paternal and maternal genetic material and on the other hand the activation of the "dormant" ovum to produce an embryo. This involves the production of different cell aggregations to form the three embryonic germ layers, which will soon engage in some well-orchestrated choreography.[39]

It is not easy to follow the developmental processes responsible for the formation of an embryo. We wonder by what magic the fertilized cell is transformed into an organism as complicated as *Homo*, consisting of billions of cells harmoniously organized in space and capable of producing organs as diverse as the heart, the limbs, the eyes, and the brain. For the biologist the challenge is to understand how cell division produces billions of cells of more than 350 different types with extremely varied functions.[40] To do this the cells must organize themselves in space according to precise instructions faithfully handed down from one generation to the next. Whatever organism we consider, the same organs are always derived from the same germ layers. Immediately after the first cell divisions the fertilized ovum is progressively transformed into a mass of cells, the morula, characterized by two germ layers, one on the outside (*ectoderm*) and the other inside (*endoderm*). The ectoderm gives rise to the external coverings of the body (skin or exoskeleton) and the nervous system. The endoderm is responsible for internal tissues, such as the digestive tract and the ducts of its associated glands like the liver and pancreas, as well as the primary germ cells, spermatocytes and oocytes.

An important stage called *gastrulation* takes place by invagination of the germ layers through an opening, the *blastopore*. In *protostomes*, the formation of the future anterior pole (the mouth) via the blastopore takes place after that of the anus. By a sort of about turn there is a reversal during evolution. In *deuterostomes*, a group to which we belong, the mouth forms first and determines the organization of the body by defining the main anteroposterior axis along which the nervous system can form around the dorsal nerve cord. Through this invagination a new germ layer is formed between the endoderm and ectoderm that we call *mesoderm*. Mesoderm produces cartilage, bone, blood cells, and muscle cells, and it participates in the formation of the heart, liver, stomach and intestines, lungs, spleen, and peritoneum. Concomitantly, under the pressure of physical forces, cell proliferation forces the three layers to move and define dorsoventral and mediolateral (left-right) axes of symmetry. In this way, through a carefully choreographed ballet with precise topological rules, the human embryo's brain can form.

Cell movements follow well-signposted routes, resulting in the formation of an embryo fitting the organizational plan of the particular species.[41] This plan is present in the egg, is independent of external influences, and is transmitted with remarkable stability from one generation to another. Animals that share the same organizational logic and the same plan constitute a phylum.[42] In *The Origin of Species* Darwin already considered the resemblance of embryos of different species as an important argument in favor of his theory of evolution. For animals of different species to resemble one another in the embryonic state, they must inevitably have shared a common ancestor, a sort of distant cousin about whom everyone in a family talks but no one really knew. This hypothesis was defended by one of Darwin's most fervent admirers, Ernst Haeckel, who later rather upset the master's ideas in formulating his theory of *recapitulation*. It can be summarized by the now famous adage that "ontogenesis recapitulates phylogenesis"; in other words, the development of an individual passes through stages of development undergone by other species of a given phylum. The different embryonic stages reproduce in an accelerated but reliable fashion the forms of adults of other species of its group.[43] Following the publication of Darwin's theories and then Haeckel's, the old concept of a single organizational plan or pattern could at last reemerge.

Recapitulation of an origin from a common ancestor could be preserved by the mysterious mechanisms of heredity. Recent molecular evidence enables us to bring this hypothesis up to date by establishing a strong link between developmental biology, on the one hand, and evolutionary science, on the other: this new biological discipline is known as "evo-devo"[44] by its proponents. In other words, this ontogenetic endeavor emphasizes the evolutionary genetics of development. The innovations that accompany the evolution of species can be studied and characterized using tools and concepts borrowed from molecular embryology. This emerging discipline owes to Haeckel the discovery of the importance of evolution in the way an embryo is formed. According to this approach, all vertebrates share a common body plan, from front to back a head, a trunk, and a tail and from top to bottom a vertebral column and an abdomen. However, for the sake of completeness, we must not ignore the constraints imposed by the environment in which the individual lives, which brings specialists to propose a new approach which they qualify as "eco-evo-devo."

We have already emphasized the origin of the nervous system as being from the most external germ layer, the ectoderm. This includes sensory cells for touch and pain, audition, smell, and taste and cells of the retina.

However, the ectoderm is also the origin of the epidermis of the skin, skin appendages, and certain glands. Only a fraction of the surface part of the ectoderm provides tissue for the central (the brain and spinal cord) and peripheral nervous system (the sensory ganglia and nerves along the spinal cord extending into the body). We now invite the reader to discover the way in which this ectoderm produces stereotyped cells of the central and peripheral nervous systems. To do this, we must discover the rules that govern the topographic organization of different embryonic structures. So, please be silent for the entry of the maestro!

The Conductor of the Orchestra

We have described how an individual's organizational plan is inscribed in the genetic material made up of molecules of DNA in our cells' nuclei. This patrimony is inherited from our parents' gametes, which fuse to form a single fertilized ovum. This DNA is the guardian of the code of the species and the individual and dictates the behavior of cells derived from the ovum according to a set of instructions that we call the *genotype*. This information allows the progressive construction of the brain, from the first appearance of the tissue outgrowth that will form the nervous system during the early stages of embryonic development to full adult maturity. This genetic program depends on a series of instructions executed at first by components of the cytoplasm of the ovum and cells derived from it as well as from the DNA in their nuclei.

During the evolution of species the genotype also guarantees the progressive formation of a variety of structures as the regions of the nervous system become anatomically more complex and control more and more varied behavior. The overall logic of the nervous system of primates, including man, is easier to understand when we consider the system not from a neurophysiological angle but from that of its evolutionary dynamics. To quote Theodosius Dobzhansky, "Nothing in biology makes sense except in the light of evolution." According to this maxim, which applies just as much to neuroscience, we can conclude that the acquisition of new neural functions in a given species is associated with the addition of new processing skills, or a capacity of adaptation to the environment, that the old functions (called primary by evolution-ists) could not provide. Darwin's aphorism that "all true classification is genealogical" comes into its full meaning when we are dealing with the history of the vertebrate nervous system.

The Story of the Embryo

Two German scientists, Hans Spemann and his student Hilde Mangold, made a significant contribution to understanding the initial stages dur-ing which the first nerve cells appear in the embryo. In 1924, using an experimental method that was revolutionary for its time, they demon-strated that a restricted region of the *blastula* (the hollow cell mass de-rived from the morula) of the newt, the *dorsal lip of the blastopore*, was capable of differentiating autonomously. This region alone of the gas-trula was responsible for the formation of the major territories of the embryo. It resembles a small split in the surface of amphibian embryos where the mesoderm penetrates to the interior of the embryo during the first reorganization of the germ layers during gastrulation. This region was identified unequivocally thanks to an experimental transplantation of a fragment of a donor embryo into a host embryo. This most famous experiment in the history of experimental embryology earned Spemann the Nobel Prize in Physiology or Medicine in 1935.

The work of Spemann and Mangold was not limited to this simple observation. In addition, they demonstrated how the transplantation of tissue taken from a gastrula and then grafted dorsally took part in the formation of the neural plate, whereas a graft placed ventrally formed the epidermis. The conclusion of their experiments was that a given region could differentiate autonomously, regardless of its new context. When they transplanted the dorsal lip of the blastopore to a ventral position (a *heterotopic* graft) in a host embryo at the same stage of development (a *homochronic* graft), the graft induced a secondary embryonic axis com-posed partly of grafted tissue but largely of host tissue. The ventral tis-sues of the host embryo were capable of transformation into, on the one hand, ectoderm to form the neural plate and, on the other hand, meso-derm, which produces the somites, precursors of muscles and vertebrae.

To explain their rather unexpected results, Spemann compared the grafted region to a true "organizer,"[45] a sort of conductor able to impose its rhythm on neighboring cells. So was born the notion of "neural induction" to describe the possibility of forming nervous tissue thanks to interaction between the dorsal lip of the blastopore and neighboring ectoderm. This discovery radically changed the way in which scientists imagined the formation of the embryonic nervous system.

This empirical step was at the origin of experimental embryology and also allowed the generalization of the concept of an organizer to all classes of vertebrates, such as fish, reptiles, birds, and mammals.[46] Following this pioneering work, numerous experimental transplantations of embryonic tissues to ectopic host regions have revealed the two principal functions that the organizer must ensure in all cases. On the one hand, it must send signals to the mesoderm, which, under the influence of a chemical gradient, is responsible for a dorsoventral polarity.[47] On the other hand, the organizer obliges the ectoderm to accept its destiny as nervous tissue. We shall discuss the nature of the signals required to assume these two functions.

Signals and Mechanisms

As Nicole Le Douarin has emphasized, until the 1930s total ignorance surrounded the nature of the signals given by the organizer.[48] Was its power of induction triggered by physical forces produced by cell movement? It was not known. Another possibility was the secretion of some unknown substance. Spemann's experiments showed that if the inductor were subjected to radical treatment such as heat, freezing, or even alcohol, it lost its power of transformation. This supported the idea that the process of neural induction depended on the presence of a chemical that such treatments denatured. It was not until the end of the Second World War that work on embryonic development brought irrefutable proof that induction by the organizer required the intervention of two key factors. One led to the appearance of essentially nervous structures (the brain and spinal cord), the other of mesodermal tissues responsible for the formation of muscles and cartilage, heart and kidneys. These two chemicals, which we now call morphogenic factors, exercise a "neutralizing" or "mesoderm-inducing" influence, depending on their respective concentrations in the embryo. These factors remained unknown until the development of molecular biology in the 1980s, which allowed access to new tools and materials.

Gothic architects of days gone by, rather than build a new church on virgin soil, often built on the ruins of a Romanesque church. It is the same with the evolution of species. Starting from existing structures, by a process of fits and starts and minor adjustments, new functions appear. A theory of *punctuated equilibrium* was proposed by two paleontologists, Stephen Jay Gould and Niles Eldredge; it reminds us that evolution consists of long periods of stable equilibrium punctuated by brief periods of major upheaval. The earlier in the tree of life that these changes occur, the more spectacular the consequences. This rule has important implications in paleontology. The further back we observe the development of an organism, the more numerous are the similarities that we can see in our primitive ancestors. Adaptive evolution since our aquatic past, for example, brought cartilage from a branchial arch, mentioned earlier, to be relocated in our jaw and in the tiny ossicles of our middle ear during the early stages of embryonic development. Let us now look at how our brain was built on old ruins.

Our New Head

The emergence of vertebrates from invertebrates was marked by the acquisition of a fundamental property that fashioned the organizational plan of all animals that inherited it. We are not thinking here of the evolution of cartilages and bones that are progressively transformed into vertebrae, as intuition might suggest.[49] The change from invertebrates to vertebrates implies much more than the simple appearance of a vertebral column.[50] It is first and foremost the choice of a life style. Vertebrates acquired the capacity to move easily to explore the world they lived in, and they became predators. This freedom essentially led to an ability to move about along a directional axis defined by the two ways of moving, forward or backward. The existence of this axis in turn defined the anteroposterior axis of the nervous system. By the intermediary of a few mutations and the duplications of some key genes, nervous structures appeared along the length of this axis. Whole families and networks of genes were "recycled" to construct new body parts associated with new systems.

This reorganization of nervous tissue led to all its major centers being located near the front of the animal, to form a head. Sensory organs and centers for decision grouped in this way can help the animal as it moves through its environment. Rather than trying to seize its food from around itself, like an amphioxus buried in the sand and obliged to feed only on particles floating past, the vertebrate can actively pursue food, a hunter after its prey. To help in this, the sensory organs became more sophisticated, permitting it to see, smell, hear, touch, and then taste its meal. This functional diversity was accompanied by the appearance of paired sensory organs symmetrically on either side of the midline. This symmetry increased the precision of perception by providing, for example, three-dimensional vision and echolocation. To render sensory and motor nerve centers more effective, they are near the mouth, where they can activate the jaws most rapidly. So, starting from a common ancestor close to modern amphioxus, mobility and sensation are the two attributes that have helped provide us with a "new head for predation," as if it evolved from the front end of a prochordal ancestor to diversify and enrich its diet. We know what happened afterward. The maximum forward development of the brain is attained in primates, with their handsome head and face, on which our species so prides itself.

In invertebrates, the motor or sensory nerve centers are spread along a nerve chain, dotted with ganglia like rosary beads. These animals are rather inactive and move little; their body is much less voluminous than that of vertebrates.[51] As they remain almost motionless, they must rely on organic particles that they ingest by filtering the water around them. To do this, a nervous system equally spread throughout the organism suffices to control such simple stereotyped behavior. This is the case, for example, of *cnidarians*, marine animals with a soft, transparent gelatinous body, the jellyfish so dreaded by holidaymakers keen to enjoy a summer swim in the sea. The jellyfish remains floating or moves around by regularly beating its bell. Its nervous system consists of a mere chain of ganglia around the edge of the bell, of which the principal activity is to maintain its rhythmic beats. Nevertheless, the nerve cells of the ganglia also receive sensory information from tactile receptors in the bell that can orient the contraction of the tentacles so that the animal can seize a prey or escape from imminent danger. So this simple, diffuse nervous system no longer merely ensures locomotion but also triggers goal-directed movements, whether toward a prey or to avoid a predator attracted by a good meal. This repertory of behavioral strategies is nevertheless not very varied. Evolution still needed to risk a few more decisive innovations.

With the first vertebrates, the agnathans, fish without jaws but with an exoskeleton of external armor,[52] a better exploration of their environment became possible. These adventurers set out to explore the world like warriors in pursuit of their enemies. Their quest would not have been possible without their nervous system's being grouped toward the front of the animal for simple reasons of efficiency and time saving. This *cephalization* allowed sensory information to trigger a rapid motor command for fight or flight.[53] The need to improve hunting performance transformed primitive sensory receptors into truly sophisticated sensory organs. Vision, hearing, and touch were especially privileged, but smell took on a very special role. Absent in amphioxus, olfaction developed in the most anterior parts of the neural tube, permitting predators to detect their prey's odor. As proof today of the primal importance of olfaction, almost the whole of the anterior nervous system of fish is utilized for the processing of olfactory information.

What is the key structure that allowed the embryo to develop and evolve a head? It is the neural crest. We mentioned this transitory vertebrate embryonic structure, which first appeared in the chordate phylum, earlier.[54] True to the principle of evolution of homeotic genes that we have already discussed, it was only after the duplication of the family of the *Hox* gene complex that the neural crest appeared and that vertebrates grew a real head. The neural crest opened the way to the invention of the modern brain. By clearing the way for the formation of the telencephalon,[55] together with the trophic action of the meninges (the membranes surrounding the central nervous system) and the solidity of the skull to ensure a good protection for the brain, the neural crest has played a key role in evolutionary innovations in vertebrates and their explosive diversification. Because the cells of the neural crest are pluripotent (that is to say, if the environmental conditions are suitable, an undifferentiated cell could theoretically furnish all the cell types produced by the neural crest), they give rise to numerous different tissues and organs. It provides cells destined for bones of the face and skull,[56] for the dorsal root ganglia of sensory systems, certain components of the walls of blood vessels, for the meninges that vascularize the nervous system, for certain glands (thyroid, carotid body, adrenal medulla), and for parts of the central and peripheral nervous systems.

As we have seen, in vertebrates the morphogenesis of the brain is closely related to that of the head. This new element opened up the world in which they could not only move but also feel, hear, and perceive face to face. The development of the facial bones provided the prototype vertebrate with a big mouth so as to snap up the first prey to face it. This predator, which preferred to devour its prey rather than suck it up, is our ancestor!

Paleontological data together with comparative morphological analysis of modern vertebrates show than the front region of the brain derived from the prosencephalon[57] has developed enormously during evolution. In vertebrates, the relative volume of the anterior brain region derived from the telencephalon increases as we move from fish through amphibians, reptiles, and birds to mammals. This expansion would not have been possible without an increased capacity for proliferation of neuronal precursors in the anterior parts of the neural tube and especially their ability to move around the embryo once they had proliferated.

Migrants on the Road

During embryogenesis, nerve cells derived from the neural tube and crest find their place after migrating from their site of origin. We should recall that there are two main types of cells in the nervous system: neurons and glia.[58] The former conduct nerve impulses to muscles, glands, and other neurons. The latter (at least the oligodendrocytes) also play an important role in increasing the speed of conduction of the impulse by producing a sheath of myelin (a lipid substance that insulates axons). During embryogenesis neurons and glia are derived from a single cell type, the *neuroblast*. Starting in the ectoderm, they migrate over long distances. That task is not easy if we consider that the human brain contains about a hundred billion or more (10^{11} to 10^{12}) neurons, gathered locally into circuits. Chemicals produced by neighboring cells are encountered all along their trajectory, allowing neuroblasts to proliferate, migrate, and survive. Overall, these cell movements follow a choreography that, as we have seen, runs from front to back, from the brain vesicles to more caudal zones. But these swarms migrating through the embryo are not there by chance. They follow well-posted paths controlled by *Hox* genes, outside which no migrants are allowed. From time to time these travelers may be invited to stop and form a cell conglomerate that will produce a ganglion of the peripheral nervous system. Others will continue under the skin to reach the fingertips to produce organs of touch. Yet others are directed toward the surface of the epidermis and become pigment cells, *melanocytes*, responsible for the color of the skin, or feathers in birds, scales in reptiles, and hairs in mammals.

In response to signals received during their migration, some neuroblasts develop outgrowths called *neurites* and become neurons. Of these multiple neurites one will become the outgoing fiber of the neuron, the *axon*, along which the electrical impulse will be transmitted, while the others will become the incoming fibers, the *dendrites*. The axon grows not from the cell body but rather from a *growth cone* at its distal end. Under the influence of homeotic genes this cone drags the axon forward in the direction in which it is growing. If its target is not reached or if the target is not active, the axon will die. So the forming nervous system is strewn with cell debris. This property of living or dying according to whether a target is reached demonstrates how the dictatorship of selector genes[59] is limited by the environment and replaced by so-called *epigenetic* mechanisms,[60] about which we shall talk in detail later. As an example of epigenesis, we might mention the case of the nature of the neurotransmitter produced by sympathetic neurons. They are produced from neuroblasts, but their fate does not depend on genes but rather on the environment in the vicinity of the neuron. One can easily verify this property by growing cells from the sympathetic chain at the base of the neck in a culture chamber. Neuroblasts pursue their natural program of differentiation into *adrenergic* neurons (neurons using noradrenaline, or norepinephrine, as their transmitter) when they are grown in isolation. However, their differentiation follows a different pathway, into *cholinergic* (acetylcholine is the transmitter) neurons, if they are cultured in the presence of glial cells or cardiac or skeletal muscle cells. These experiments illustrate the variability of neuronal precursors, which are able to alter their fate and modify their chemical content under the influence of the environment. Independently of their origin in the neuraxis, these precursors are able to respond to chemical signals from embryonic tissues in order to differentiate into ganglia, plexuses, or endocrine cells according to their position in the embryo. In embryology nothing is written in advance.

How to Build a Vertebrate

The three embryonic germ layers (ectoderm, mesoderm, and endoderm) are already in place during early embryonic development. During later stages organs are formed, dependent on very varied cellular functions including proliferation, migration, differentiation, and programmed cell death.[61] All these functions depend on the intervention of proteins inside the cell, on its membrane, or secreted into the extracellular matrix between the cells, part of which provides the *basement membrane* to support epithelial cells.[62] The *Hox* genes that permit the expression of these key proteins are called *architect* genes. They are at work during the final phases of embryogenesis and control the formation of different parts of the body. Because they have the vital function of placing organs where they should be and outlining their contours precisely, the activity of architect genes is finely regulated. To do this other genes fulfill the role of accelerating or suppressing their activity. Like the *regulators* defined very broadly by François Jacob and Jacques Monod, these selector genes act in appropriate cells at the appropriate time and in mutual coordination, to imprint their identity on the different parts of the body.

The existence of a neuronal zootype implies that the action of selector genes, architect genes, and homeodomain genes be coordinated in time and space. In this way we perceive that the principal function of *Hox* genes is not only to build the foundations, whether skin (outside) or bone (inside), but also to ensure the establishment of communication systems to link all parts of the embryo, fulfilled by the vascular and nervous systems. A muscle in a given part of the body will be labeled unambiguously by *Hox* genes. This label will facilitate its identification by a nerve terminal seeking to contact it,[63] which will simply have to read the instruction left by the *Hox* gene in order to establish a permanent connection with its target. In vertebrates the embryo is formed progressively, from the front, before the real body segments, the paired somites derived from mesoderm, can be added successively from front to back as the embryo lengthens. The somites form the vertebrae and muscles innervated by spinal nerves. *Hox* genes are progressively expressed as somites are added, both in the neural tube and in the somites.

In the vertebrate embryo, the most anterior end of the nervous system consists of three vesicles, the *rhombencephalon* or hindbrain, the *mesencephalon* or midbrain, and the *prosencephalon* or forebrain. In mammals the forebrain is further subdivided into three structures: the *optic vesicle*, which forms the retina and optic nerve; the *diencephalon*, which becomes the thalamus and hypothalamus; and the *telencephalon*, which gives rise to the olfactory bulb, cerebral cortex, and basal ganglia, as well as white matter in the form of fiber bundles such as the internal capsule and corpus callosum. The fibers of the corpus callosum (figures 1.2, 1.4) play a crucial role in cerebral function: they ensure communication between the right and left cerebral hemispheres. In chapter 5 we shall come back to the deficits a patient suffers if these fibers between the two hemispheres are damaged.

Beyond its organization into three primitive vesicles, a segmental organization is visible from the beginning of the fourth week until the end of the fifth in the embryonic human brain. These are temporary narrow swellings called *neuromeres*. In the fifth week we can distinguish a telencephalic neuromere, four diencephalic, two mesencephalic, and eight rhombencephalic. The rhombencephalon consists of repetitive juxtaposed units called rhombomeres. Each unit has its own "nucleus," a mass of neuronal cell bodies of which the axons converge to form major cranial nerves. For example, the facial nerve (cranial nerve VII) and the acoustic nerve (cranial nerve VIII) are derived from cell bodies situated respectively in rhombomeres r4 and r5.

Studies of mutations in mice demonstrate that the trajectory of nerve fibers coming from the rhombencephalon remain under the strict control of *Hox* genes.[64] Even more remarkable is that these studies show that homeosis involving a region can affect nerve pathways situated in neighboring areas. To interpret these unexpected results, Alain Prochiantz and his team at the Collège de France found evidence that a homeodomain protein (Otx2) produced in the retina was able to cross several synaptic relays, highly selectively, enter nerve cells of the visual cortex, and reach their nucleus, where these proteins exercise their power.[65] This demonstration of a homeoprotein behaving like a messenger protein and acting at a distance to exercise a morphogenetic action turned a new leaf in developmental biology. We described earlier the extent to which homeoproteins are extremely well conserved across species and the major role they play during development. This recent discovery suggests that these messenger proteins, which are also produced throughout the lifetime of an individual, could be exploited to favor the repair of certain diminished neural functions.

In summary, we have seen how each individual fashions the marvelous shape of its brain from embryo to adult. At the summit of human brain development, from about ten to sixteen weeks after conception, around 250,000 neurons are born every minute to reach the sum of a hundred billion neurons existing at birth. From twelve months of age, while still incapable of holding a conversation, babies show signs of surprising cognitive faculties. For example, psychologists in Marseille recently showed that neonates can handle statistics to forecast probable events in their environment.[66] Jean Piaget proposed that such probabilistic inferences were not possible before seven years, the age of reason! Nevertheless, when one-year-old babies were shown videos of containers inside which were placed objects of different shapes and colors, bouncing randomly, a sort of anticipation by the child was observed. When a container with five blue balls and a red cube was shown and then hidden, the babies showed great surprise if the red cube bounced out of the container but not if a blue ball emerged. They expected a blue ball to come out because that represented the highest probability. Even more surprising was that these same babies were able to perform a probabilistic analysis of the situation by considering not only the respective proportions of the different objects (blue balls versus red cube) but also their place in the container (at the bottom rather than the side). From a very early age our babies' brains are thus capable of performances that we are only just beginning to suspect. At about six years of age the brain reaches some 90 percent of its adult size. However, as we shall see later, we must not assume that that is the whole story—far from it. The establishment of shape is characteristic for a given species and is a result of the evolutionary history of the individual with its landmark genetic innovations. To conclude this chapter on the progressive growth of the human brain, we might compare it to those other masterpieces, the cathedrals: like them it remains unfinished, ages without ever reaching maturity, and is the object of constant repair and restoration. So now, if you will, please follow the guide, who is ready to take us behind the scenes of this permanent construction site.

3

THE MASTERPIECE

If I lost my brain I wouldn't even notice.
—ANONYMOUS

IT IS PARADOXICAL THAT THIS ORGAN, which makes us what we are, should have remained unrecognized for so long. Indeed, we had to wait until the eighteenth century to discover its external and internal structure, thanks to a multitude of cadavers dissected in anatomical amphitheaters and to advances in microscopy. As to the study of its function, that depended on ridding physiology of religious prejudice, the advent of electricity, and the study of pathological anatomy.

The study of key stages in embryonic development, which we discussed in chapter 2, demonstrates that the vertebrate nervous system is a mosaic where phylogenetically old structures exist side-by-side with newer additions. This is illustrated by the structure of the cerebral cortex, which we divide into two parts according to their phylogenetic origin: an older *allocortex* (from the Greek *allos*, meaning "other"), which represents hardly 10 percent of the cortical mantle, and the more recent *neocortex*[1] (from the Greek *neos*, meaning "new"), which forms most of the rest.

To appreciate the adult brain's capacity for *plasticity* it is essential to have an idea of the structure and topography of both the peripheral and central nervous system. In a guided tour of the brain[2] one might exclaim, "How can a cathedral like that be housed in the 1,500 cc (91.5 cu in) of the human skull? A total mystery! It's not surprising that the brain continues to inspire a sort of sacred admiration mixed with suspicion. Its discovery and exploration came after those of America. Before, this *terra incognita* was a subject of speculation and superstition. Almost untouchable. We may still hesitate even to let our children study the brain. Too complicated, we argue. Yet is it reasonable not to let them learn the function of an instrument that allows us to act, to love, to know? It's a little as if one went to Egypt without visiting the pyramids, as if one crossed Rome avoiding the Coliseum, or Athens without looking at the Acropolis. This why I propose a sort of guided tour inside the brain."

Without more ado, let's visit this cathedral, where we can distinguish not only a division between *central* (the nave of the cathedral) and *peripheral* (its arches and buttresses) but also two major divisions within the peripheral nervous system: the *somatic* and the *autonomic* systems. The former deals with the relations of the organism with its external environment. Its nerves transmit information from various sensory receptors to the brain. The latter intervenes in the regulation of major vital functions, maintaining the equilibrium of our internal environment by coordinating such essential activities as digestion, the circulation of the blood, respiration, and the secretion of hormones. Depending on circumstances, the autonomic nervous system plays the role of both firefighter and pyromaniac: activation of its *sympathetic* system helps prepare the organism for action (physical or mental) by stimulating major vital functions, and the *parasympathetic* system does the opposite, saving energy by slowing down the same vital functions.

The peripheral nervous system[3] remains relatively resistant to an individual's experience, whereas the central nervous system remains extraordinarily flexible, quick to respond to the lessons of the past, changing its structure to modify its functions. Indeed, it is this ability that makes the analogy of the brain with a computer wholly inadequate.[4] With its hundred billion processors and a million billion connections, the brain is incomparable to anything from the world of information technology. Whether considering processors or software, people too often commit the error of wanting to apply unconditionally the metaphor of the computer and the brain.

Of all the natural or artificial objects in the universe, the human brain is certainly the most complex. This complexity is expressed, on the one

hand, by the juxtaposition of very different component parts, each with more or less specific functions,[5] and, on the other hand, by the sheer number of nerve cells: about a hundred billion neurons and even more glial cells.[6] We have to add to this complexity the huge number of contacts (synapses) between nerve cells.[7] On average, a neuron receives (and makes) about ten thousand contacts, with which it continuously exchanges electrical and chemical signals with other cells to collect, process, and store meaningful information to ensure that the subject is properly adapted to its environment.

The grayish-yellow mass of the brain, weighing on average some 1,500 grams (3.3 lbs.), consists of two distinct regions. The cerebral cortex, which covers most of its surface, is relatively little influenced by genes but is constantly being remodeled by subjective experience. Its nerve networks are unstable, "plastic," and profoundly influenced by factors within and outside the body, which provide epigenetic regulation, as we described in chapter 2. On the contrary, deeper brain regions react structurally much less to the environment and to subjective experience. These regions remain stable, are genetically determined, and have an ancient evolutionary history. We shall see how these subcortical regions enable an individual to adapt to its environment by managing the totality of its major vital functions. The great novelty of the human brain is unquestionably the diversity of form of its neural circuits that each individual can shape to its own needs. We now know that at the cortical level the same genotype can produce a very large number of phenotypes[8] or a single phenotype that experience will modify under the influence of epigenetic regulation. This dialectic between genetics and epigenetics demonstrates the great extent to which adaptive behavior in the adult depends on developmental mechanisms.

Because the central nervous system is alone able to integrate and manage information from the outside world, it is through it that the organism can adapt to a changing environment, a process that we shall call individuation.[9] To consider individuation as the result of an adaptive cognitive process (like perception, language, memory, or conscience) is to admit that subjective experience can be expressed in the function of one's own neural circuits. So now let's lift the skullcap over this exquisite machine that is able to change its integral parts to increase its efficiency.

The Cortical Brain

How can we fit the 2.5 square meters (27 sq ft) of our unfolded cerebral cortex into our small skull? If we compare skulls with the brains they contain, we discover, as might be expected, that a mold of the inside of a skull relates to the shape of the cerebral hemispheres. The skull increases its interior volume as the brain increases in volume and changes shape. Phylogenetic study of the evolution of the skull demonstrates that as the hemispheres increase in size from the front they pivot on their central axis, at the level of the pituitary stalk, and then develop further back. The orientation of the Sylvian fissure reflects movement of the temporal lobe as it gradually follows the backward tipping of the head, which is so important in hominins.[10]

To visualize the brain we merely need to saw round the skull cap and remove it by detaching the bone from the underlying meningeal membrane called the *dura mater*, which forms a sort of shell around the brain. We then open the dura, cut the brainstem where it merges with the spinal cord, and remove the brain from its box. The brain remains covered by its other two delicate meninges: we have in our hands something like a large pinkish fruit with a pearly luster, whose wildly convoluted cortex for long defied description. The largest and most obvious part of the brain is formed by the two oval-shaped *hemispheres* (figure 1.1). Their irregularly flattened inferior surface sits on the base of the skull and partly covers the cerebellum posteriorly, separated from it by a fibrous extension of the dura. The left and right hemispheres are separated by a deep *longitudinal fissure* but are linked by massive *commissures*: the largest is the *corpus callosum*, a thick band of white matter containing some 200 million axons forming a bridge between the hemispheres (figure 1.2). This highway, which transmits information from one hemisphere to the other in a few hundred milliseconds, allows us to experience consciousness as a single whole. As we mentioned in chapter 1, if it is divided, two minds, rather than a single one, must learn to cohabit in the same skull.

The surface of the hemispheres is covered by numerous *sulci*, which delimit *gyri*. Deep sulci called *fissures* separate the *lobes*. Each lobe contains

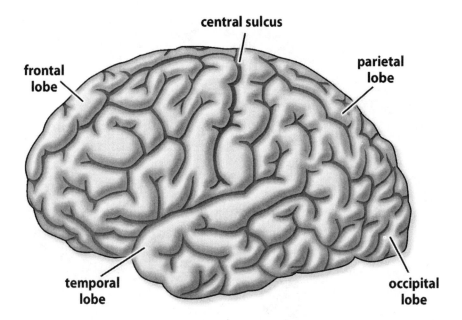

FIGURE 3.1. The four lobes.

a number of gyri, and on the surface we can distinguish four lobes in each hemisphere: *frontal, parietal, occipital,* and *temporal* (figure 3.1), and buried deeply and out of sight are the *insula* and *cingulate gyrus.*[11] In functional terms the frontal lobe plays an important role in an individual's behavior, especially in the planning and control of activity and sociability. The parietal lobe facilitates representation and exploration of space. This feature allows the brain to control gesture and have an overall familiarity with the body in three dimensions, thanks to input from cutaneous receptors for touch, temperature, and pain. Situated below the parietal cortex, the temporal lobe is the seat of integration of various sensory modalities such as hearing, taste, and smell. It is also responsible for understanding the meaning of words and visual memory. Finally, at the extreme posterior pole of the hemisphere, the occipital cortex decodes visual information, analyzing shape, color, and movement.

The cortex itself consists of gray matter three to four millimeters (0.12 to 0.16 in) thick. Its histology is complex and varies from one

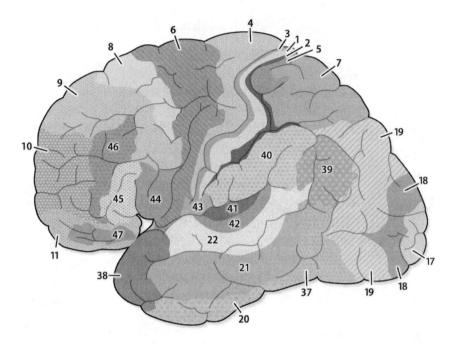

FIGURE 3.2. Brodmann's 1909 lateral view of the human brain, with his numbers for different functional areas.

region to another. The cell bodies and fibers are organized in layers. At the beginning of the twentieth century, using microscopy, Korbinian Brodmann distinguished different areas, which he numbered (figure 3.2). For example, area 4 corresponds to the primary motor cortex situated anterior to the central sulcus.[12] One of its layers contains *giant pyramidal* neurons. Pyramidal neurons give rise to myelinated fibers, sheathed in *myelin*, a lipid insulating substance derived from glia, which serves to increase the speed of conduction of a nerve impulse. It gives the typical pearly white color to the brain's white matter. A nerve impulse arising in the primary motor cortex causes muscles to contract on the opposite side of the body. Parts of area 4 innervating individual muscles are distributed so as to form a map of the body, so-called *somatotopy*, like a grotesque human form, the *homunculus* (figure 3.3),[13] of which the size of each part is proportional to the density of innervation of its muscles. For example, there is an enormous hand with a gigantic thumb, a large face, and an

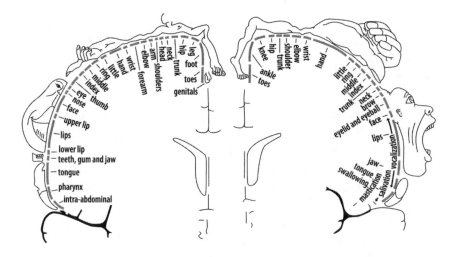

FIGURE 3.3. The sensory (left) and motor (right) homunculus. Redrawn from W. Penfield and T. Rasmussen, *The Cerebral Cortex of Man* (New York: MacMillan, 1950).

overrepresentation of the tongue and muscles of mastication. On the contrary, the foot and leg are thin, and the trunk is underrepresented. There is also a sensory homunculus in the parietal lobe behind the central sulcus. In chapter 6 we shall describe the amazing capacity of the homunculi to modify their shapes in cases of accidents such as the loss of a limb or after grafting a hand.

In addition to this horizontal organization, by which cortical layers are superimposed like layers of a "mille-feuille" pastry, the cortex and its connections form a series of *columns*. This vertical organization permits complementary functional activity in motor and sensory areas. Indeed, these columns behave like processing modules situated between the input and output of data signals.[14]

In the next chapter, we shall see that each individual does not perceive its environment in the same way because of the extreme sensitivity of cortical areas to acquired experience. Thanks to the existence of such reorganization, everyone can develop his own personality according to the dual influence of history (*ontogenesis*) and species (*phylogenesis*). These two forms of evolutionary pressure are joined by cultural influences from interaction with other individuals to enable man to live uniquely and free!

The neural pathways that project to and from the cortex cross the midline so that each hemisphere connects with the opposite (*contra-lateral*) side of the body. This anatomical phenomenon of crossing, or *decussation*, of nerve fibers has been known for a long time. Even Hippocrates noted that a wound on the left side of the head caused convulsions on the victim's right side. The adaptive value of such decussation still remains enigmatic among scientists. The right cortex controls movement and sensation of the left half of the body and receives input from the left half of visual space; the left cortex deals with the right half of the body and visual field. However, neural pathways and cerebral projections are not always equal on both sides, as we saw in chapter 1. Certain functions are *lateralized*, in that the two hemispheres may not receive the same amount of information, one side being seen as *dominant* over the other. For example, language is controlled by the left hemisphere in right-handers and the right in left-handers. But even here nothing is absolute. If a right-hander loses his right hand, the right hemisphere (still connected to the functional left hand) will take over. As we can see, political leanings are not relevant where brain science is concerned.

What Is Inside a Walnut?

The subcortical brain contains various *nuclei* of gray matter (figures 1.3, 1.4). They include the basal ganglia of the *telencephalon* (caudate nucleus, putamen, globus pallidus, substantia nigra, subthalamic nucleus) and a small mass of gray matter separate from the cortex of the hippocampal gyrus, the *amygdala*. The *diencephalon* is situated medially in the deep part of the brain and contains the thalamus and, in its floor, the hypothalamus.

The thalamus (from the Greek for "inner chamber") is the major portal of entry of pathways to the cortex. All sensory afferents, with the notable exception of olfaction, relay in the thalamus before reaching the cortical sensory areas. Amazingly, each sensory modality has its own relay nucleus in the thalamus, like a post box. For example, the medial geniculate nucleus receives information from the inner ear and passes it to the auditory cortex, whereas the lateral geniculate nucleus receives information from the retina and sends its axons to the visual cortex. Although most information received by the thalamus comes from sensory systems, it also relays other neural activity, such as from the cerebellum, the basal ganglia, or the temporal lobes, all of which establish reciprocal connections with the thalamus. Importantly, the sensory relay nuclei of the thalamus do not only send fibers to the cortex but also receive powerful descending connections from the cortex. For example, in the *thalamic reticular nuclei* descending corticothalamic projections control thalamic activity. This regulation is achieved by means of reticular neurons producing a powerful inhibition of thalamic output to the cortex. Through this filter, sensory information reaches its cortical target, or not, producing conscience experience, or not. Some scientists, like Rodolfo Llinàs of New York University, consider that the convergence of information in the thalamus, together with the existence of reciprocal connections between the thalamus and cortex, provide a basis for consciousness. Nowadays most neurobiological models of consciousness emphasize activity in thalamocortical circuits.[15]

Inferior to the thalamus a visitor will find the *hypothalamus*, where visceral mechanisms that take part in homeostasis of the internal environment are regulated.[16] Here we also manage our needs, our desires, our pleasure, and our suffering. This region is a veritable vegetative powerhouse, and it plays a fundamental role in the integration of somatic, autonomic, and endocrine function, for the hypothalamus receives information from the viscera and responds directly to changes in the internal environment. For example, the long-term control of fat metabolism is now better understood since the discovery of a hormone, *leptin*, which is secreted by fat cells and acts directly on the hypothalamus to inhibit food uptake. The hypothalamus is a storehouse where the objects of our desires are hoarded, but it also has the capacity to act on the rest of the organism through the endocrine and autonomic nervous systems, of whose activity it controls. An outgrowth from the floor of the hypothalamus, the *posterior pituitary gland*, liberates two major hormones, *oxytocin* and *vasopressin*, which are implicated in the control of childbirth, milk formation, and blood pressure. Other parts of the hypothalamus control secretions of the *anterior* pituitary. In fact, the hypothalamus contributes to a central representation of the body by maintaining an up-to-date register of an individual's state in three domains, corporal, extracorporal, and temporal, through which it expresses its subjectivity.[17] We shall return to the degree of flexibility of this structure when we discuss functional changes of affect.

Subcortical centers include the diencephalon, parts of the limbic system, the basal ganglia, and the brainstem. These regions are not characterized by neurons in layers as in the cortex but rather as cell groups or *nuclei*. Some are large, like the basal ganglia deep in each hemisphere or the amygdala inside the temporal lobe. Others are smaller and may be pigmented, like the *substantia nigra* (black) and the *locus coeruleus* (blue) in the brainstem. Of course, like all classification, this division of the mammalian brain into cortical and subcortical regions has its limits. Although it may be anatomically correct, it does not always represent a functional reality. For example, the *limbic system* contains both cortical and subcortical elements. We have to accept that neural circuits do not always recognize the arbitrary borders placed by scientists anxious to attach their name to a nervous structure to ensure their posterity.

The two hemispheres of the brain are placed on top of the brainstem, which merges with the spinal cord. Along the cord run descending pathways, carrying electrical signals from the brain to motor neurons, and ascending pathways, carrying sensory information from the body and outside world. The brainstem contains nuclei that occupy the space between the fiber tracts running to and from the brain and the rest of the body. These nuclei participate in major autonomic functions and enable integration of signals from the body. Notably, they deal with visuomotor signals, those concerning hearing and balance, facial motor control and sensation, the mouth and throat, and the respiratory system and heart. The structures of the brainstem and various deep, medial parts of the brain intervene in a complex fashion in what we conventionally call states of consciousness: minimal consciousness, sleeping, dreaming, waking, and attention. Lesions of these regions cause very varied pathological conditions, such as deep coma or states in which the subject totally loses the use of his muscles while preserving his consciousness of the world, becoming a prisoner of his own body. In chapter 6 we shall return to a case that has become famous, that of the author of *The Diving Bell and the Butterfly*, who suffered a cerebrovascular accident in the brainstem.

This brief tour of the brainstem brings us next to the *reticular formation*, a system for *integration* of convergent information from ascending pathways and for divergent *activation*, keeping the brain awake. It is responsible for maintaining the cerebral cortex "switched on" so that perception, voluntary activity, and even thought can occur optimally. This formation is important in controlling attention and intervenes in cycles of waking and sleeping. It contributes to homeostasis of the internal environment through numerous hierarchical subsystems in the brainstem. Finally, we should note that the association of the reticular formation with the striatum forms an anatomical substrate for major adaptive behavior of an individual and a species, as we shall discuss in detail in subsequent chapters.

While we are on this tour we must pause to consider the basal ganglia. They form a group of nuclei at the base of the telencephalon deep to the anterior parts of the lateral ventricles. There are three subdivisions: the *caudate nucleus*, the *putamen*, and the *globus pallidus*. The caudate and putamen constitute the *neostriatum*, for both are phylogenetically recent. The globus pallidus is the *paleostriatum*. The striatum has a function in learning, especially of motor tasks (so-called *procedural* memory). When we learn to ride a bicycle it is thanks to the learning capacity of the basal ganglia that we progress and quickly become expert cyclists. The function of the basal ganglia in the initiation and maintenance of a motor command is clearly illustrated when they are diminished in Parkinson disease. Patients have great difficulty in initiating movements, and they perform them with characteristic slowness and tremor. The dorsal part of the striatum is mainly concerned with motor functions, but the ventral part is intimately associated with affect. It receives pathways that use dopamine as their transmitter: these dopaminergic fibers control the balance between desire and aversion, producing affective states that we experience as pleasure or suffering.

For a long time it was thought that pleasure was related to the utilitarian value of behavior. This view has changed today toward a biological view that dissociates pleasure from satisfaction of a need. Once again we owe this conceptual change to our discovery that the brain is flexible or plastic. This new view depends, among other things, on the results of experiments on self-stimulation, which demonstrated the dynamic character of function in neural circuits. These experiments consisted of inserting an electrode in a rat's brain to let the animal stimulate certain brain regions. From the very first series of electrical stimulations delivered to the right region by pressing a lever, the animal would not stop pressing it, even hundreds of times an hour, neglecting its fundamental needs such as eating and drinking. The animal became a slave to its own behavior, creating a dependence that was the behavioral equivalent of neurological *addiction*.[18] Study of the circuits that had received these stimulations showed radical changes of form and function at every organizational level examined by the experimenters—neurotransmitter receptors, synapses, neurons, and circuits—even over a very long time.[19] The brain region involved is related to pathways from the brainstem where dopamine is synthesized to be released by the *nucleus accumbens* of the basal striatum, connected to the frontal cortex. The outer *shell* of this nucleus manages our emotions through its links with the amygdala and the limbic system. Because the inner *core* is implicated in motor control, the accumbens acts as an important interface between desire and action. By detecting the quantities of dopamine released and regulating the brain's selective contact with the outside world, it plays a vital role in controlling our affect.

Desire (or, technically, *wanting*) is above all desire for reward related to a need, of which the satisfaction brings pleasure. It expresses a vital bodily need for provisions such as water, nutrition, and energy. The end result of desire should therefore be pleasure, and the choice of a behavior to engage should be dictated by the pleasure procured. However, this definition cannot be taken as a general rule. It is possible that self-stimulation may activate wanting and pleasure in parallel and not in series. In this case wanting would not always be linked to need, and pleasure would not necessarily result from the satisfaction of this need. Simultaneous activation of the wanting-pleasure pair would explain the insatiable character of self-stimulation and especially the "gratuitous" nature of the pleasure that can sometimes be observed. Detailed analysis of the psychostimulatory effect of so-called reward drugs (cocaine or amphetamine, for example) enables us to distinguish two categories of behavior: that depending on motivation (linked to wanting) and that producing the pleasure associated with consumption of the drug. The first is pure *wanting*, and it depends on dopamine. The second, *liking*, is related to consumption and satisfaction of a need and depends on pathways that release other chemicals, *endorphins*. These two neurotransmitter systems operate synergistically. The dopaminergic system related to wanting is strongly associated with the characteristics of the desired object and explains the appetite-like aspect of some behavior related to our approach to and appropriation of objects. The endorphins are endogenous opioids related to consumption and to physiological and metabolic consequences of satisfying a need (consumptive behavior).

We should emphasize that every situation that creates a particular affective state, for instance a pleasant one, creates an opposite state in parallel, in this case an unpleasant one. This so-called *opponent-process* mechanism is based on Claude Bernard's homeostatic equilibrium, taken up by Walter Cannon, who spoke of the *wisdom of the body*.[20] The opponent-process theory, which takes into account the influences of the internal and external environment, was first described in 1974 by Richard Solomon of the University of Pennsylvania and John Corbit of Brown University, to explain a number of particularly intriguing emotional phenomena.[21] The explanations they proposed enable us to understand the powerful emotional changes that occur according to a time frame specific to a given situation. For example, Solomon and Corbit cite the case of parachutists who often experience terror when they first jump but, once that is over, feel a different, equally intense emotion of exuberance or even exhilaration. The case of lovers who experience great passion when they meet and are plunged into profound sadness when their loved one is absent illustrates another situation in which happiness precedes suffering. The two examples demonstrate that there exists not only a temporality in affective responses but also that their polarity can flip. This reverse process is derived as a counterreaction, develops after a certain delay, and persists long after the end of the stimulus. That the process of reaction and counterreaction are out of phase is at the origin of a rebound effect. This is the famous state of well-being experienced by the marathon runner when he crosses the finishing line, his body wracked with pain inflicted by his prolonged effort. This late effect increases with repetition of the stimuli and tends to overshadow the primary process that triggered it, whereupon more and more powerful stimuli are necessary as the subject becomes tolerant, and the secondary effect increases with the suffering.[22] Now we can see that plasticity of the adult brain does not always lead to paradise—far from it.

Limbo of the Mind

According to the theoretical neuroscientist Paul MacLean, who in 1958 divided the brain into three parts on evolutionary grounds, as we saw in chapter 1, the limbic system corresponds to a paleomammalian brain, the seat of our motivation and emotion (figure 1.4). It can respond to incoming information after comparison with a register of past information. This deep part of the brain intervenes in processing emotion, learning, and memory. The limbic system consists of a complex of nerve centers and their linking pathways, which border (*limbus* means "border" in Latin) the neocortex. It forms part of what in 1868 Paul Broca termed the *grand lobe limbique*, around the base of the neocortex. Two paired symmetrical structures, situated medially in the brain, are at the heart of emotional processing: the *hippocampus* and the *amygdala*, which we discussed briefly in chapter 1. The former manages an individual's personal space and his relational maps with the world. It is also strongly implicated in memory formation. As such, the hippocampus is the model of predilection for scientists seeking a biological explanation for the mystery of memory. The second region, the amygdala, so named because of its almond-like shape, is buried deep in the temporal lobe. It serves to recognize emotions, especially fear, on others' faces, to express it, and to establish associations in situations that in themselves are not congruent. Thanks to these structures an individual learns to associate pleasure or suffering with an object or situation and to estimate the intensity and valence (hedonic or aversive) of a stimulus.

The hippocampus, so named because of its shape, which recalls that of a seahorse, receives input from practically all regions of the cerebral cortex through a sequence of successive relays. Anatomical studies show that the hippocampus forms a loop, like a drive sweeping up to a hotel, where one can both enter and leave. Its role can be summed up as comparing the state of the world and its affective value. Nervous activity in hippocampal circuits is rhythmic, at a periodicity of 10 to 200 milliseconds (5 to 100 hertz). This oscillatory electrical activity seems to play an important role in memory formation. It is present during learning and appears strongly during dreaming, incidentally illustrating a probable link between memory and dreams. In view of its connections with the cingulate cortex and the mamillary bodies of the hypothalamus, the hippocampus also plays a major role in emotional aspects of memory.

Appetite, sexual behavior (and consummatory behavior in general), and defense strategies developed during evolution depend on interactions between the limbic system and the brainstem. They form a system linked to internal organs and to the endocrine and autonomic systems, autonomously regulating cardiac and respiratory rhythms, sweating, digestion, and body cycles associated with sleep and sexual activity. The circuits of this brainstem-limbic complex form relatively slowly reacting feedback loops (from a few seconds to a few months). These have no precise topography in that they lack a functional map. These loops are specific to meet the individual's internal needs but not to adjust to multiple signals from the outside world. They appear early in evolution to meet the fundamental needs of the organism in a constantly changing environment. The amygdala is the center of activity. It nestles in the medial part of the temporal lobe and has a double relationship with the hippocampus: one is direct, the other indirect, though the *stria terminalis*. Activation of the amygdala produces different effects depending on which part of the autonomic system, sympathetic or parasympathetic, is involved. On the other hand, its ablation, like that of most parts of the limbic system, reduces fear and anxiety. The subdivision of the amygdala proposed by Alf Brodal in 1947 into corticomedial and basolateral is important, for it confirms different observations following stimulation or ablation of these regions.

The limbic system consists of two components responsible for pleasant or unpleasant effects. The *septal nuclei*, the *medial forebrain bundle*, and the *hypothalamus* can produce pleasant effects, and the emotions that accompany them often have a strong sexual connotation. On the contrary, activation of the amygdala and its efferents, in part the stria terminalis, produces reactions of rejection and disgust. Overactivity of the amygdala can have disastrous social and economic consequences for an individual; it can contribute, for example, to the production of a future homicidal delinquent. The consequences of lesions of the amygdala demonstrate the importance of this deep nucleus. When amgydalectomy was still practiced for epilepsy, a reduction in reactions of fear and aggressiveness was observed, with an almost total incapacity to attribute an affective significance to sensory messages. Such apparently heartless individuals remained indifferent to photographs with even the strongest emotional content, such as dismembered bodies. Neurologists began to study patients with Urbach-Wiethe disease, a very rare genetic syndrome in which there can be calcification of the amygdala. These patients are incapable of recognizing sadness or fear on other human faces. They might caress an enormous tarantula or grab a snake without the slightest fear.

So far, to describe the origin of affect we have voluntarily limited ourselves to a rather brief description of subcortical regions, in particular the limbic system and its related structures. It is now necessary to emphasize the importance of the cognitive dimension and the relationship between affect and consciousness. For that let's follow the guide into the nave of our magnificent cathedral.

The Vertebrate: An Emancipated Animal!

As opposed to that of an invertebrate, the body of a vertebrate is soon emancipated from the genetic patrimony inherited from its parents. From the first stages of embryogenesis, epigenesis is at work. This allows individuals to develop differently depending on what happens after the egg stage, the end result of which is the formation of a unique organism. In other words, the production of an individual depends on its history and escapes the dictatorship of its genes. In vertebrates we do indeed find a basic plan to follow, but the final construction may differ from the architect's project. The liberated animal can thus make its presence felt in the world.

In invertebrates, such as insects, although their cognitive and sensory capacities are highly developed such that they are able to detect and discriminate wide ranges of shades or subtle odors imperceptible to man, such tasks are performed with no possibility of improvisation. Insects are provided with effective automatisms, but the absence of plasticity in the nervous system makes them incapable of following rules that are outside the normal register for their species. In brief, their ingenuity has its limits. Nevertheless, some invertebrates seem capable of learning. The nematode *Caenorhabditis elegans*, very highly regarded by neurobiologists as a model for studying behavioral plasticity, is capable of remembering odors associated with food or punishment. The sea slug *Aplysia* is another example, famous among neurophysiologists since Eric Kandel of New York's Columbia University demonstrated the great flexibility of its neural circuits when learning tasks.

In vertebrates, the central nervous system adopts its familiar overall form: a tube protected by a vertebral column, capped anteriorly by three basic swellings themselves protected by the skull. As we saw earlier, the formation of the neural crest offers not only the possibility of harmonizing the organism's activities via the peripheral nervous system but also a greater capacity to adapt to the outside world thanks to neural circuits being grouped anteriorly. For example, the peripheral nervous system can control body heat as a function of outside temperature by innervating blood vessels that dilate or constrict according to what is needed. "Cold-blooded" animals can therefore be joined by "warm-blooded" ones that can maintain their body temperature constant. In man this same system has become a means of interpersonal communication by signaling the affect of an individual, for example turning white with fear or red with anger. Major communication functions have been facilitated by the emergence of the neural crest, whether involving exchange of information between organs, between an individual and its environment, or most importantly between individuals. Once again, in vertebrates the true meaning of confrontation with others is seen with the development of a head.

Nevertheless, the invention of the neural crest does not only lead to increased freedom for vertebrates. This freedom has its price. With the neural crest, systems of reward and punishment also appear. These systems are the basis of pleasure and suffering and positive and negative reinforcement. They influence behavior to allow approach or flight depending on the affective state of an individual. Whereas invertebrates are in general incapable of emotion, vertebrates experience the relationship of their bodies with the world through their brain. Thus a sort of liberty, a sort of slack in determinism, seems to regulate human behavior via affect systems. More precisely, man is characterized by the existence of an intersubjective state, a central dynamic state. This state is like a musical trio interpreted by the body—performed by the hormones, the nervous system, and the immune system. It is the world within which the body functions and which it perceives, and with which it interacts. And there is the dimension of time, the distant past that we carry in the genes of our species, and our own personal history. In the end, no individual vertebrates have the same past, the same temporal dimension, even if their genetic baggage is comparable. The vertebrate can indeed cry out, "Long live the difference!"

4

THE WORKSHOP OF THE BRAIN

What is life?—Life—that is: continually shedding something that wants to die.

—FRIEDRICH WILHELM NIETZSCHE,
THE GAY SCIENCE (1887)

THE CONSERVATION OF LIFE on earth is just as exacting as that of energy: both are continually degrading, and all living organisms are destined to sink inevitably into oblivion. Life is only possible thanks to the repair, restoration, and reconstruction of degraded organs, until the workshop closes at death. Many invertebrates are able to regenerate a missing limb or organ, including their nervous system. The so-called higher species, especially man, seem to have lost this remarkable potential. On the other hand, they have available numerous means of recuperation to allow certain handicaps to be overcome, at least partially. If this recuperation is indeed possible it is because, contrary to what we for long believed, the construction of even the adult brain is never totally finished. Indeed, will it ever be? The great innovation of vertebrates is to have kept a nervous system that preserves embryonic properties even in the adult. Thanks to this extended ontogenesis, an individual can benefit from plasticity to allow changes in the form of neurons, the replacement of synapses, or the addition or elimination of neurons. The preserved dynamism of neural circuits enables adaptation to be expressed at the individual level rather than by selection of clones, as in invertebrates. People often imagine a brain fixed in structure and function. Not at all. Our cathedral is animated and evolving. We shall now look at a dynamic brain.

Tissue Regeneration

Regeneration is the ability of living matter (cells, tissues, organs, organisms, ecosystems) to reconstitute themselves after amputation or partial destruction. Restoration of a lost organ is not only structural: it means also recovering its function, totally or partially. When this concerns the nervous system it forms part of the arsenal of *neuroplasticity*, which plays a major role in optimal neural functioning. Recently propelled to the forefront of the medioscientific bandwagon, the existence in the adult of proliferative zones where new neurons are born continuously illustrates an extreme case of regeneration, to which we shall return later.

The ability of some animals to regenerate parts of limbs or organs is in itself a fascinating subject of study.[1] Even more remarkable is the ability to reconstruct a damaged nervous system. The fortuitous discovery by Abraham Trembley in 1740 that the freshwater hydra could be cut in two and regenerate the whole or part of its body, so as to make two, laid the foundations of the discipline that was later called regenerative medicine. Destined by his parents to become a minister of the church, the young Abraham decided to reject a spiritual future by launching into a brilliant scientific career, first in mathematics, then in natural science. As a budding scientist he found some hydras in muddy water and wondered if they should be classified as animals or plants. To solve the dilemma, the custom at that time was to cut the mysterious organism in two, like King Solomon trying to discover the truth. If the organism did not survive after being cut it was considered to be an animal. To his great surprise not only did the two parts not die, but Trembley found that he could obtain up to fifty hydras from one!

Sadly, man has not inherited their ability. For a long time scientists have tried to understand the origin of this mystery. Two hypotheses have been put forward. One suggests that an ancestor of the metazoans was able to regenerate but lost the ability during evolution (a vestigial property). The alternative hypothesis states that this ability was absent in the ancestor of the metazoans but appeared, in various forms, during evolution (an emergent property). According to this hypothesis, regeneration was an important adaptive trait that natural selection retained. Research to confirm or reject either of these hypotheses has not yet provided a complete answer. We now know that the spinal cord can regenerate after amputation of the tail in aquatic anamniotes (fish and amphibians) and reptiles. Studies in fish also demonstrate that their brains can regenerate. On the other hand, self-repair is rarer in birds and rarer still in mammals. These observations seem to support the "vestigial" theory, according to which regeneration is blocked, lost, or reduced during evolution. Clearly, the significance of these experimental data needs critical evaluation. In particular we must not forget that the potential for recovery of nervous function depends on the region affected and also the context in which the lesion occurred.

When we try to demonstrate the ability of nervous tissue to regenerate, the age of the individual is also an important parameter to be considered. Amphibians, birds, and mammals all show partial recovery from a spinal cord lesion at the larval or embryonic stage but lose the ability by adulthood. Even if the potential is still there, notably thanks to the presence of neuronal stem cells in certain nerve centers, it seems that tissue environment determines the regenerative capacity of adult tissues. The loss of the ability to retain immature physical characteristics in the adult (a trait called pedomorphism) seems to be responsible for the scarcity of regenerative processes in higher vertebrates. Because regeneration in frogs, chickens, and rodents depends on the stage they have reached, it is probable that it is a property inherited from a common ancestor.

In the Tree of Life,[2] the forms of adaptation differ according to whether we consider the branch that bears the invertebrates or that of the vertebrates. Even though we learn a lot about the way in which we construct our nervous system by studying how the fly or the worm do it, the rules that permit us to develop adaptive strategies are radically different. In arthropods most adaptive changes are at the genetic level, via clone selection. In brief, there is no place for individuation in invertebrates, for the genome dictates the future of the species. On the contrary, in vertebrates, and more so in mammals, each individual is different from its brethren because of environmental pressure, although the characteristics of the population and the species are preserved through intersubjective communication and sexual reproduction. The richness of information exchange allows adaptation, and individual variability, through an individual's history.

In consequence, the nervous system of a vertebrate at time t is not quite the same as at time $t + \Delta t$: it has simply evolved. The intensity of synaptic activity, their number, the total number of nerve cells in a given structure, and the spatial and temporal organization of networks will have changed. This constant adaptation of adults to their environment, dependent on their experience, is ensured by the continuous expression of developmental genes that enable the evolution of the species and the formation of major neural structures (cortex, cerebellum, spinal cord) but that also permit the nervous system to be ever adjustable and flexible in form and function. Stephen Jay Gould, in his famous work *Ontogeny and Phylogeny* (1977), already mentioned this possible mechanism for evolution. In particular, he put forward the idea that a modification of the chronology of embryological or ontogenetic development could coincide with the adaptation of a species without implying significant phenotypic changes. As we have seen, this concept applies perfectly to the vertebrate brain. Alain Prochiantz even considers this phenomenon as a major element for adding an extra dimension to evolution.[3]

New Neurons in the Adult

The vocal center of birds and the olfactory bulb and hippocampus of mammals are brain structures that permanently produce new nerve cells, whatever their age. From research on zebra fish, canaries, mice, but also on man came the idea of a new form of cerebral plasticity based on adult neurogenesis, plasticity about which much still remains to discover, from its mechanisms to its functions.

The discovery in 1992 in the adult mouse of stem cells able to produce new neurons in the adult brain held out immense hope.[4] However, their use for the treatment of neurological disease still remains dependent on a better understanding of their basic biology. Nevertheless, large strides have been made. Until quite recently we considered the adult brain as an organ with no capacity to regenerate and inevitably condemned to lose its most precious elements, the neurons. This fixist view was profoundly shaken by the discovery that the adult brain could produce new cells. Some fifty years ago Joseph Altman proposed for the first time the

possibility of a proliferation of nerve cells in the adult rat.[5] Curiously this potential aroused only limited interest at the time, reinforced by the absence of proof that the new cells that Altman saw were really neurons.

We had to wait until the 1980s for the notion of new neuronal production to be rehabilitated by research on the canary brain. Since then, this bird's brain has remained a cutting-edge model in neurobiology. When Fernando Nottebohm of Rockefeller University, New York, started an ambitious program in the 1970s to identify the neurological mechanisms of learning, little did he suspect the surprises in store, especially the extent of the morphological changes he was to discover in the brains of his charming canaries. First he identified the groups of neurons involved in motor learning and singing. He then demonstrated that the song system contained two circuits under the control of a nucleus, the *high vocal center* (HVC). On the one hand was a motor pathway controlling the execution of the song, on the other a regulatory loop integrating auditory information and allowing the bird to compare the song he was singing (by listening to it) and a song he had memorized.

One of the team's greatest surprises was the discovery in 1983 that new neurons are produced in the canary's HVC. Unexpectedly they discovered a process of renewal of nerve cells, whatever the age of the individual. They watched as new neurons continuously replaced old ones that had degenerated. This renewal seemed specific, concerning exclusively a subpopulation of HVC neurons, those of the motor pathway. The other subpopulation remained unchanged. So the question arose as to the functional significance of this specific production of neurons. Then they showed that there was a link between the rhythm of the seasons and the intensity of neurogenesis and between neurogenesis and annual cycles of performing and learning by the male of his song to seduce the female. The male canary's mating song is only performed in a stable stereotyped form in the spring. He stops singing in late summer. After that, in the fall, comes a period characterized by the execution of improvised song, then at the end of winter a phase of new learning. Finally in the spring a new song is introduced, different from that of the previous year, thanks to the incorporation, suppression, or modification of certain characteristic syllables. Remarkably, Nottebohm and his collaborators found two peaks of neuronal death in the HVC, one in August, the other in January, precisely the two periods during which the canary's song becomes unstable and deteriorates. Equally remarkably, this increased cell death precedes the maximum rate of production of new neurons, which is highest in October and March. Taken together, these ethological and morphological observations revealed the relationship between the acquisition of new syllables and the replacement of old neurons by new.

Later, the same team discovered an explanation for these periodic variations. They demonstrated the involvement of testosterone in the process. In the canary the production of this hormone is regulated by the daily light level, which increases significantly in spring. Testosterone modulates the expression of a factor needed for neurons to survive. In the fall, when the concentration of testosterone drops the expression of this survival factor diminishes, and neuronal death increases. Inversely, when the hormone increases in spring the factor increases, and the survival of newly produced neurons is favored. So the annual change in the song repertory is a direct response to seasonal changes in the canary's environment, which regulate the number of new neurons. A beautiful example of a brain custom-made for seduction! A few years after this research on the canary, when the ideas had matured further, the first evidence emerged of the renewal of cerebral neurons in adult rodents, primates, and then man. Such a change in paradigm would certainly not have displeased Thomas Kuhn.[6]

We now know that several parts of the mammalian brain are able to produce and host new neurons. They include the hippocampus, a key structure for spatial memory that we already mentioned in the previous chapter, and the first relay in the olfactory system, the olfactory bulb. In both cases new neurons are formed from a germinative zone of neuronal stem cells, the subgranular zone of the dentate gyrus of the hippocampus and the subventricular zone bordering the cerebral ventricles in the anterior part of the brain. Stem cells proliferate, migrate, and are transformed into true neurons just as they do during the construction of the embryonic nervous system. Recently, stem cells have also been identified in the association cortex of the macaque, an important region for storing active memories and one fundamental to thought processes. Although it seems probable that brain regions concerned in adult neurogenesis are associated with memory and learning, the question of the physiological significance of renewal of neurons remains: we shall return to it later.

In spite of the growing interest in studying adult neurogenesis, may questions remain unanswered. How can the integration of new neurons in a mature network operate without upsetting the stability of preexisting circuits and the maintenance of neuronal processing? One commonly accepted hypotheses is the recapitulation of embryonic processes in the adult. Neurogenesis, embryonic or adult, depends on various cellular mechanisms, such as proliferation, migration, differentiation, and cell death. But can we state, as some do, that embryonic mechanisms are similar to those operating in the adult? Given their permanent renewal of neurons, the hippocampus and the olfactory system are well suited to attempt an answer to that question. Work on the hippocampus indicates that adult neurogenesis is continuous with the molecular and cellular programs that control embryonic brain formation. The situation in the olfactory bulb might be different. The regions where bulb neuroblasts are produced and integrated are contiguous in the embryo (as is the case for the adult hippocampus) but far apart in the adult. In consequence the molecular factors that act locally to control the different stages of proliferation, maturation, and integration of future neurons are different in nature, depending on the stage under consideration. In so far as these genetic and epigenetic factors differ, embryonic neurogenesis may well turn out to be very different from that in the adult. In the adult, the ability to produce new neurons for the olfactory bulb might reveal mechanisms unique to the adult brain.

The existence of adult neurogenesis shows that a capacity for adaptation in the nervous system is not only at the level of variations in synaptic activity or morphological changes in neurites. By escaping from genetic determinism, these adaptive processes in the adult brain play a crucial role in fitting an individual to its environment. Of course, an individual's adaptation is not independent of genetic adaptation because it depends intimately on genetically programmed developmental strategies. However, by allowing a better adaptive response by an individual than by the species as a whole, secondary neurogenesis privileges individual experience over phylogenesis.[7] Thanks to the brain, which is so important in the transition from invertebrates to vertebrates, the range of adaptive possibilities progressively favors the individual at the expense of the species. The consequences of this cerebral plasticity on individuation are obvious: they enable the history of an individual to be written in the form of permanent morphological and functional alterations in neural networks. In other words, an individual's brain is able to record the chronicle of its life and the story of its impassioned relationships with the world.

These adaptive processes, derived from a close interrelationship between the genetics of the species and the history of the individual, are essential for the emergence of cognitive and behavioral functions. The more a subject is stimulated, the more it develops various epigenetic constructions. That is true for infants and adolescents but also for adults, thanks to cerebral plasticity. During development certain sensitive periods are determinant for the acquisition of a new function, one that could be lost irrevocably, depending on circumstances, if not acquired during a certain window in time.[8] Such periods of opportunity constitute phases during which learning is extremely effective. However, learning that does not take place during this period can be acquired later, throughout the subject's life, but it is simply more costly in terms of time and effort. This property of the nervous system is most marked in the early years of life. For sensory stimuli, certain emotional experiences, and some complex cognitive functions like calculation and musical comprehension, the sensitive periods are relatively short, up to a few years in man. It is also true for other animals. For example if a kitten's eye is closed for a week at a specific time, it loses stereoscopic vision. Almost all its visual cortical neurons cease to respond to the eye that was closed. In mammals, this feature of the nervous system during the early stages of development may seem odd in that they are not much exposed to the environment. Other mental faculties, for example reading and vocabulary, do not adhere to this notion of a critical period and can be learned throughout life. Thus there exists a form of learning dependent on experience, a learning characterized by a sensitive period and is consequently limited in time. Acquiring vision or the sounds of language belongs to the first type of learning. On the other hand, the acquisition of lexical or semantic learning, which can take place throughout life, belongs to a second category. Now we shall give more precise details about this very sensitive period in youth that precedes full maturity of the adult.

Juvenile Plasticity

For some species the time needed to reach maturity is short. For example, invertebrates in the course of their very brief ontogeny develop characteristic phenotypic traits selected by conditions imposed phylogenetically on the species before the individual existed. This rapid phase of development represents an advantage because it allows the young animal to survive more effectively in a hostile world. So scarcely ten days after birth *Drosophila* can already reproduce. On the other hand, other species choose the opposite strategy and favor a long period of development. In this case, a large number of young are sacrificed to allow a few to survive. Another survival strategy for animals with a long ontogenesis consists of prolonged parental care to protect young in an environment where all sorts of predators are rife. In mammals, for which sexual maturity may take decades, the second option is chosen. This period of intense relationships between newborn and parents offers the young exchanges with their environment which shape them as they develop. The same is true for man, who remains relatively fragile for the first twenty years of life, dependent on the family group, and whose neotenic brain,[9] only reaching adulthood relatively late, is fed by social interaction.[10] Biological order bows to psychic order, and social and cultural transmission become primordial for the development of the brain. Factors such as diversity of diet or complexity of parental care appear as important actors in molding the brain. This type of flexible ontogenesis that precedes brain maturity is found in most species that enjoy much phenotypic plasticity.[11] This juvenile plasticity does not have precise temporal limits. Depending on the function in question, the initial phase of this flexible ontogenesis remains the same for a given species, like the kitten that can only hear from the fifth day after birth and opens its eyes a few days later, but the end of this period remains vague: it is sometimes prolonged beyond childhood.

The existence of juvenile plasticity led the pioneer of cognitive development, Jean Piaget, to suggest that the intellectual development of the child was not progressive but moved in steps corresponding to different critical stages in the acquisition of mental faculties. Piaget distinguished four major stages in cognitive development: a sensorimotor stage from birth to two years; a preoperational stage from two to seven years; a concrete operational stage from seven to eleven years; and finally a formal operational stage, or hypothetico-deductive stage, from eleven years. Each stage was characterized by a distinct cognitive plan and by the level of complexity of mental activity. In man, old age is a period of cognitive senescence in which learning capacity is reduced, unlike in the young subject, in whom learning is optimal during these four periods. In other words, during a human lifetime there are times for learning and other times for retaining already acquired information. Youth is characterized by a period during which the subject learns quickly and consequently is highly innovative, but "old" people are individuals who, through the stability of their behavior, guarantee social cohesion by contributing moral and social references. This dichotomy, where learning and stabilization are the central pillars of memory, recalls the rules of machine learning in robots, a favorite aspect of artificial intelligence when it seeks to develop machines able to learn from past experience. We learn that a machine can only learn if training is balanced by phases of stabilization to allow stimuli to be categorized; this is an indispensable stage for the long-term storage of information.

Studies on adolescents confirm that their brain has not yet reached full maturity. Even if the brain of a ten-year-old has already 90 percent of the volume of an adult, the work is still far from complete. Brain imaging shows that the brain continues to grow until the age of twenty or thirty. Its neurons become more myelinated, and the adolescent brain undergoes considerable structural modifications under the influence of hormonal spurts, even after puberty. A second wave of synapse formation and pruning culminates in adolescence, between twelve and twenty years of age after the first wave in infancy. Some regions are especially affected by these secondary changes. The first is the striatum, which helps regulate desire systems: this may explain high-risk behavior in adolescents looking for a quick reward. The second is the epiphyseal region (the pineal gland and related brain structures), where melatonin is secreted, which helps synchronize our activities with daylight. An adolescent produces melatonin later in the day than an adult and has greater difficulty in sleeping before midnight. The prefrontal cortex is also a privileged target where there can be a major reorganization of brain circuits in the adolescent. This region is a controller that constantly anticipates, plans, evaluates alternatives, makes us conscious of negative (the right prefrontal cortex) or positive emotions (the left), and ensures executive functions.[12] Nevertheless, the dominant function of this zealous controller is exemplified by the remarkable inhibition it can exert. If it cares to, it can suddenly stop the programming of a movement before it is executed. Thanks to its action on the amygdala we can restrain joy or hold back a tear in public. So the prefrontal cortex is the support of human sociability and can always keep its calm even in the face of the greatest injustice. But in the adolescent it is still immature: it is not able to impart that emotional stability which, in principle, characterizes adult wisdom, and so we see unstable and sometimes dangerous behavior in some adolescents.

This immaturity and instability of cognitive function in the adolescent also poses the problem of the efficiency of school selection practiced on subjects whose psychic abilities change from month to month. We shall return in the chapter dealing with the sick brain to the great vulnerability of the adolescent brain to mental illness or addiction. In the United States more than 1 percent of very young infants, 2 percent of primary-school students, and 5 percent of adolescents suffer from depression, with twice as many girls affected as boys; biological factors of strictly organic origin are certainly indistinguishably mixed with social factors, as with everything concerning the brain.

Connectional Plasticity in the Adult Brain

The idea according to which our perception of the outside world is represented in the brain topographically, as a cognitive map, is quite ancient. Permanent interaction of a live subject with its environment is necessary, for organisms constitute what is called in thermodynamics "open" systems: they only exist thanks to a constant flow of matter, energy, and information.[13] Therefore one primordial function that the brain must fulfill is to ensure the analysis, filtering, and integration of sensory input to construct a mental representation of the outside world adapted to an animal's specific activities. The brain can exercise this function because it is able to represent general or durable features of the world thanks to a more or less stable configuration of its myriads of synaptic connections. This configuration of carefully regulated connections, which is able to evolve with time, determines the way in which the brain inscribes its environment in its own circuits. This representation is contained, in part, in the particular configuration of the hundred thousand billion synapses in our brain. It is a consequence of the law of large numbers, thanks to cephalization, the grouping of tens of billions of neurons in our skull. From this complex society of highly interconnected nerve cells is born the faculty to learn difficult tasks. In this way neural networks are able to recognize a letter, the features of a familiar face, or a sound.

This aptitude is not innate, for no biological attribute, such as DNA, could suffice to contain all the necessary information for the construction of our mental representations. The structure of neural networks, then, is derived in part from contact with the environment, and thus learning is indispensable for their elaboration. It can be achieved by a double mechanism: connections between neurons might at first be random and redundant, but later only connections between simultaneously active neurons are retained (Hebb's principle). This is a selection phase, which is accompanied by the elimination of weak contacts. As we already mentioned, Jean-Pierre Changeux and Antoine Danchin elaborated a theoretical model of the possible evolution of nerve connections depending on the age and experience of a subject. This model postulates that there exists in the young brain a large number of connections, a given neuron having synaptic contacts with hundreds if not thousands of other neurons. Under the influence of passing time and experience only functional connections are stabilized; the others degenerate.[14]

The majority of neural networks learn thanks to their ability to classify, categorize, generalize, memorize, but also to forget. Quite rightly, Plato said that to learn was to remember what one had forgotten. During learning and during childhood development synaptic connections are modified according to one's needs, to produce assemblies of neurons that function together on the same task, giving rise to functional maps. The demonstration that the adult brain can produce new nerve cells allows us to glimpse the evolution of the contours of these maps through learning.

Neural networks are thus dynamic ensembles that adapt according to activity and depending on intercellular factors. The topographic representation of sensory perception in the mammalian cortex is able to reorganize itself according to changes imposed by the environment. Today we have broad experimental proof that neurons have an eminently dynamic structure and function that enable them to adapt to change both internal or environmental.[15] In 1949, Donald Hebb suggested that the maturation of a nerve pathway depended on the amount of information transiting through it. Since then many studies have validated and extended that principle, showing that factors modulating the environment or afferent neuronal activity can modify the anatomical and functional organization of pathways. Recent neurobiological studies demonstrate that interaction between electrical activity triggered by the periphery and the genetic program significantly modulates the essential properties of neural networks required for coding sensory information.[16] From this point of view every individual is unique.

Olfaction: The Prototype of Epigenesis

We can only access the physical world that surrounds us through our sense organs. Philosophers have for long reflected on the relations within the world beyond our perception, that is to say relations between thought and object. Different philosophies hold different views as to whether perception is a simple association of elementary sensations or a whole, organized in advance. This is not the place to reopen the controversy, but we can say that this philosophical question is now open to experimental investigation. Since it has been studied scientifically by specialists, conjecture has given way to experimentation. Psychologists and neurobiologists, wishing to study the *real* world, take a *subjective* view, in the true sense of "related to the subject."[17]

To survive, most animal species must adapt their behavioral strategies to variations in their environment. This remarkable behavioral adaptation originates in often complex mechanisms that intervene at the sensory receptors (where raw information is collected), at the integrative stage (where information from different receptors is filtered, processed, and coded), and finally at the perceptual stage (where the mental picture of the received information is formed). These mechanisms, which depend on receptors and integrative centers, enable us to trigger behavior that is fundamental for life and the evolution of the species as well as for reproduction, flight from danger, and the search for food. Such behavioral responses have in common that they are broadly dependent on sensory systems in which olfaction plays an essential, even vital, role.

Olfaction is at the forefront of sensory modalities.[18] The sense of smell begins with the first breath of the newborn. Are not smell pathways in the brain the earliest to form during ontogenesis? This precocious sense may allow the fetus to recognize the chemical signature of its mother. At birth olfaction provides the baby with its initial confrontation with the world. All the fundamental behavior established during fetal development is profoundly associated with olfactory signals. Later, this sense is used to recognize and select nutrients, to detect toxic substances and rotten food, to attract and recognize sexual partners, and to establish parental and social bonds. The great novelty of olfaction is its capacity to renew certain of its neurons, which only live a few weeks. This permanent renewal is possible thanks to the presence of neural stem cells buried in the sense organ (the olfactory epithelium lining the walls of our nasal cavities) or in the deepest parts of the brain, in the walls of the lateral ventricles. In such a dynamic context, how does a rose continue to smell like a rose if our olfactory neurons are constantly replaced? And how is our memory for smells so robust if our neurons die after learning? Indeed, we know that smell above any other sense is tied to memory.[19] In this context we might mention the role of olfactory memories in the genesis of basic behavior in a species and that the so-called *rhinencephalic* structures are at the center of memory in our brain.[20]

To try to understand the complex and still rather mysterious function of the olfactory system, it is first useful to distinguish two levels of organization. The first is in the periphery, the olfactory epithelium of the nasal cavities. This is the sensory organ for smell, a chemical receptor in engineering terms. Odorous molecules stimulate receptor neurons directly. The second level is the olfactory bulb, a small forward extension of the olfactory paleocortex (figure 1.1). It processes sensory signals from the receptor neurons to transmit them to cortical and subcortical centers

that together deal with perception and recognition of more or less sweet odors. As we have seen, these two regions, peripheral and central, have the very rare faculty of adult neurogenesis,[21] thanks to the presence of neural stem cells, with their remarkable property of being able to change into any other cell type after division. In the olfactory bulb these stem cells produce glia: oligodendrocytes and astrocytes. These cellular dynamics pose de facto the question of the role of new cells in the mature brain. They also open new clinical perspectives by offering novel solutions for the repair of lesioned regions. The presence of neural stem cells indubitably represents new hope for the treatment of certain neurological pathology, including vascular or traumatic accidents: we shall return to this later.

Today we have little difficulty admitting that adult neurogenesis contributes to the long-term progressive adjustment of the function of neural circuits, but numerous questions persist. Can we distinguish, compare, or classify embryonic *versus* adult neuronal production? How does a newly produced adult neuron migrate to its target? How does it choose its destiny? We shall try to answer some of these questions using those structures of the olfactory system concerned in adult neurogenesis: the sensory organ and the first cerebral relay.

The nose, the vanguard of the face, is an admirable chemical instrument. In the nasal fossae we find a mucous membrane in which sensory neurons react to chemical elements. In man this organ has a surface area of two to three square centimeters (0.3 to 0.47 sq in) on the upper lateral and medial walls of the nasal cavity. In addition to sensory neurons it also contains stem cells, the basal cells, which can divide and form new sensory neurons. Because the olfactory epithelium receptors are directed toward the outside world, they are subject to infections or toxic agents, such as pesticides and heavy metals. In such extremely adverse conditions they only live a few weeks. The presence of stem cells[22] in the epithelium thus seems important for repairing the sense organ as it is so often damaged.

Let us now leave the periphery, enter the brain, and pause at the olfactory bulb, where new cells are produced from the subventricular zone in the walls of the lateral ventricles, deep in the brain. Recent advances in the study of the architecture, organization, and function of this germinative zone show that it derives its unique properties from the local cellular and molecular environment in which the stem cells are nested. We call this special environment a *niche*, a term borrowed from ecology to stress the importance of interactions between different cell types, membrane molecules, soluble substances, and other extracellular material present close to the stem cells, the ensemble of which constitutes a veritable ecosystem. During development these niches exist temporarily in the embryo but persist in the adult, although we do not really understand the reason for this pedomorphic characteristic. To reach the olfactory bulb cells from the walls of the human lateral ventricles must migrate over several centimeters. The journey lasts several days, during which new cells acquire their status as true neurons. Recent results point to two fundamental attributes of neurogenic niches in the adult brain. On the one hand, they contain glial cells (astrocytes) able to change from a resting state every month or so and divide rapidly (every day) to provide neuronal precursors. On the other hand, these niches must fulfill two other missions: provide a *permissive* environment for cell division and exercise an *instructive* function so that the fate of the new neurons is adapted to what is needed.

Our team has shown that stimulating the brain facilitates production of new neurons. The idea seemed far-fetched a few years ago, yet a few weeks of cerebral or physical stimulation sufficed to double the number of new neurons produced in the rodent brain.[23] These newly arrived cells in the olfactory bulb process sensory information differently from neurons made in the embryo.[24] This postnatal neuronal production enables the personal experience of an individual to make its mark repeatedly on neural circuits. In the adult, new neurons may arrive at certain special periods when memory needs reinforcing. In summary, new neurogenesis is probably an additional adaptive mechanism to permit constant renewal of an individual outside the famous critical periods. We might wonder how new neurons are used in a system as dynamic as olfaction, what might be their function, and in what context they might express themselves. It is clear that the parts of the brain concerned in adult neurogenesis are closely associated with memory and learning, but the question of the physiological significance of this neuronal renewal remains a mystery. Studies on mutant mice demonstrated a deficit in discrimination between two similar odors when the recruitment of new neurons was interrupted.[25] On the contrary, animals raised in an enriched sensory environment manifested a more developed and robust olfactory memory than weakly stimulated controls.[26] There is also evidence for social implications of adult neurogenesis in the mouse showing that the production of new neurons in the olfactory bulb was stimulated after mating, during the first week of gestation, then during the first week of lactation.[27] This increase in neuronal recruitment could be important for maternal care of neonates: under the influence of hormones such as prolactin, new neurons reaching the olfactory bulb might increase empathic behavior in the mother, who becomes ready to respond to the slightest alarm signal from her babies.

The anthropology of human olfaction is marked by three principal characteristics: the rich olfactory palette (more than 10,000 molecules recognized on average, and up to 40,000 in certain professionals such as perfume designers and enologists), instantaneous triggering of pleasant or unpleasant emotions depending on the odor, and rapid evocation of memories and images. Odors, with their vague cognitive content (it is often difficult to name an odor) but their marked emotional tone, remain for us a direct link to identifying the elements around us. Today we can add to the three attributes mentioned the faculty of the olfactory brain to produce and integrate new nerve cells continuously, as we have just seen. This ability is an extreme example of plasticity, under the control of personal experience and interaction with the environment, and contributes to the individuation of the subject. We wager that future research will enlighten us on the relative importance of the different mechanisms that contribute to the complexity and individuality of a human being.

When the Brain Goes Wrong

Every day almost 80 percent of our cerebral activity depends on visual signals. So how is our activity affected when vision goes missing? Neurologists observe that the area where vision is processed, the primary visual cortex of the occipital lobe, remains active, for it is "recycled" for other sensory modalities. This piracy, to borrow a term more often associated with informatics now that freight less often follows maritime channels, illustrates the phenomenon of *intermodal plasticity*. According to this principle someone born blind might use his visual cortex, recruited by tactile sense, to read Braille. By using transcranial magnetic stimulation to inactivate the cortex temporarily, it has been shown that activation of the visual cortex remains indispensable to read Braille. In other persons blind from childhood, listening to music can activate the visual cortex. The team led by Maurice Ptito at the University of Montreal School of Optometry attempted to make the most of this piracy. They produced a visual prosthesis that used the tongue rather as one uses the fingers for Braille. A hybrid visuotactile device transformed an image captured by a digital camera into electrotactile stimulations decoded by the tongue. Because the tongue is the most highly innervated organ in the body (remember the sensory homunculus we described earlier), the information perceived produced an image five times more precise than one obtained by stimulating the skin.

These examples of rehabilitation demonstrate the plasticity of primary sensory areas, which can be redirected from their original function for another modality. To explain the nature of this recuperation experiments were conducted on rodents. They showed that in mice deprived of vision very early direct connections were established between the visual, auditory, and somatosensory cortex that were not present in visually intact mice. These discoveries show that cortical functions are more like those of a Swiss Army knife than those of a brand-new computer.

When the Cortex Takes Over a Grafted Limb

From numerous more or less spectacular clinical cases neurology has shown us that the adult brain is capable of reconfiguring itself in the face of sometimes extreme situations. This is true in experiments involving virtual reality. These have shown that a subject's appropriation of his own body can be deceived. In 1998, Matthew Botvinick and Jonathan Cohen, from Carnegie Mellon University and the University of Pittsburgh, developed an experimental method to show that it was possible to readjust consciousness of the spatial unity of the body and the psyche.[28] Using a clever device, they observed that subjects could take over ownership of a rubber hand although it was not theirs. The subject observed an artificial hand while his own hand was hidden. He looked at the artificial hand while the rubber hand and his own hidden hand were stroked with paintbrushes. Subjects experienced the illusion that they felt the touch on the rubber hand. So, vision influences our touch sensation and even dominates it, and our experience of our body is the expression of multisensory integration. What an excellent example of piracy, in this case by vision, which manages to convince touch to accept a virtual limb that does not belong to it! In fact, the conscious experience that we have of our body stems more from our access to multimodal areas in the right parietal cortex than it does to access to primary somatosensory areas.

Incidentally, this experiment also shows that the body unit is not rigid, as witnessed by numerous cases of *out-of-body experience* (OBE) related to imminent death.[29] Such illusions correspond to an extreme, disconcerting body experience that, depending on the case and the subject, can be pleasant or frightening. An OBE gives the impression of seeing the world from an elevated disincarnated position and of perceiving an image of one's own body beyond its physical limits.[30] To feel oneself within the physical limits of one's body and to perceive the world from this very egocentric position[31] depends on the convergence and proper integration of vestibular signals with visual and somatosensory information.[32] If sensory information is not properly integrated, the unity between self and body is interrupted.[33] The experiment by Botvinick and Cohen shows that in the case of multisensory conflict concerning the body the visual illusion dominates over other sensory modalities. The participation of the vestibular system in the perception of the environment and the attribution of parts of the body to oneself is illustrated by clinical cases in which electrical stimulation is given during neurosurgery, for example, in epileptic patients resistant to treatment by medication. We know that low-intensity stimulation at the temporoparietal border induces strong vestibular sensations of rotation and at higher intensities illusions of falling, leaving the body, or disincarnation, which cease immediately after the stimulation is stopped. These manipulations of body unity offer a scientific interpretation for OBE and other sensations perceived during the experience of imminent death, explanations that were long relegated to the realms of mysticism.

Meditation and Cerebral Plasticity

Many contemplative traditions consider meditation as the result of particular cerebral activity, but, conversely, is there a causal relationship between regular meditation and cerebral activity? Does the mere fact of thinking, meditating, or contemplating modify connections in our brain? Most scientists, fans of rationalism, of course thought that the answer was simply no. However, over the last few years we have seen research into this problem, showing that the question is far from trivial and at least deserves to be asked. In this context the American Society for Neuroscience invited the Dalai Lama to their annual meeting in Washington in 2005. The declared object of his visit was to promote fruitful debate on meditation and contemporary knowledge on brain function. For the Dalai Lama neuroscience has made enormous progress in fundamental knowledge over the last thirty years, especially in the fields of motivation, attention, and emotion. It was thus time to build a bridge between the scientific discipline and meditation or contemplation.[34] Meditation, he believed, was simply a technical means to improve attention and take full control of one's affective states. During his talk he declared: "The meeting of modern neuroscience and Buddhist contemplative discipline, therefore, could lead to the possibility of studying the impact of intentional mental activity on the brain circuits that have been identified as critical for specific mental processes." Francisco Varela of the French National Institute of Health and Medical Research reached the same conclusion when he and his collaborators attempted to bring together neuroscience, phenomenology, and Buddhism. They wondered, for example, what science and Buddhism taught us about perceptual illusions, dreams, knowledge, or memory. By seeking to understand the relationships between consciousness, body, and brain—and how modern research can upset such notions—Varela ultimately was asking the question of the nature of the mind according to modern science and Buddhist experience.[35]

We must conclude that this reconciliation is no longer taboo, as we can see from recent results in the field. In Richard Davidson's laboratory at the University of Wisconsin, Madison, experiments were made to identify changes in brain activity during and after meditation.[36] Tests were made by electroencephalography[37] on eight Buddhist monks and a group of ten volunteers trained in meditation for a week. The training consisted of meditating on compassion and love. Two of the volunteers— and all the monks—showed increased gamma activity (around 25 to 40 hertz) during meditation. When they stopped meditating, the waves returned to their initial value in the volunteers, whereas the monks showed increased gamma activity for some tens of minutes afterward. We should emphasize that the monks had some ten thousand hours of practice in meditating on compassion. A notable difference during the meditation was also that the gamma activity was stronger in the monks during meditation than in the same brain regions of all the volunteers. These results illustrate another, still little-studied aspect of cerebral plasticity triggered by thought or meditation.

Another example of this new interest can be seen in the work of Jon Kabat-Zinn of the Center for Mindfulness in Medicine in Worcester. He is interested in *mindfulness-based stress reduction* (MBSR). The members of his laboratory study MBSR in the context of long-lasting disease accompanied by chronic pain. They have shown how meditation can deactivate self-referring neural circuits solicited by powerful pain. At the same time, their work has contributed to a better understanding of the mind-body interface.[38]

We could mention as a last example work using functional magnetic resonance imaging (fMRI) on practicing adepts of Vipassana meditation. This research shows the important role of the *insula*,[39] a cortical structure involved in interoceptive awareness, that is sensation from deep viscera. Thanks to this structure we perceive our bodily sensations and remain informed about the state of our viscera and thence our levels of stress and temperament. It is probable that such results may one day find applications in the human clinical field, such as attention deficit or anxiety. In the United States there are more than two hundred hospitals where meditation is practiced daily as an adjunct to pharmacological treatment and to manage anxiety in patients with terminal cancer. It is high time for neuroscience to begin to unveil the mystery of the effect of meditation on brain function.

Plasticity in *Homo cultiorus*

What if all forms of plasticity we have described only exist because of culture? For Merlin Donald of Queen's University at Kingston and the Case Western Reserve University, to belong to a culture and communicate in a common language is to be part of a network of collective knowledge and interpretation, in other words, a cognitive community. "Symbolic thought and language are inherently network phenomena," the origin of language being "in cognitive communities, in the interconnected and distributed activity of many brains."[40] Starting from fundamental observations from anthropology, neuropsychology, primatology, and archeology, Donald seeks to establish a general theory of the origins of human cognition and their influences on modern human cognitive machinery. His work gives us food for thought:

> Humans have super-plastic brains. A developing human brain is a sort of snowballing cognitive Leviathan that adapts to everything and anything close to it. . . . But our super-plasticity would not have much, if any, adaptive value outside the context of a highly unpredictable cultural world. In most species, extreme plasticity would probably prove to be a liability, leading to instability.[41]

In the end a very plastic brain might be a weakness with no adaptive value if the subject lives in a stable, predictable world, the opposite to that of *Homo sapiens*. On the contrary, the creative and cultural spaces produced by human mental activity would be the real driving force of cerebral plasticity. As Edgar Morin said in 1973, "we saw *Homo sapiens* take a majestic leap away from nature and produce through his great intelligence, technology, language, society, and culture; we also see nature, society, intelligence, technology, language, and culture co-producing *Homo sapiens* by a process lasting millions of years."[42]

5

THE BRAIN UNDER REPAIR

Ask my arse, my head hurts.
—POPULAR SAYING

AS WE HAVE SEEN, a new head is the hallmark of vertebrates. We have every reason to be proud of it and to want to protect it, by a hat against sunstroke, or by horns or tusks against adversaries who might attack it. Indeed, the head is the first thing to be attacked in combat, a sad privilege shared with the heart, also indispensable for life.

Man is an inquisitive animal, and we can easily conceive that since his origins he has sought to know what was in that skull of his, which rang hollow and broke when you hit it. Well before civilization, prehistoric man practiced trepanation, an act still difficult to interpret. The Rosetta Stone of brain history consists of two famous documents: the Edwin Smith papyrus and that of Eberts. They are the earliest-known documents containing clinical observations on wounds to the brain and attempts to treat them by surgery. However, until the Renaissance descriptions of the damaged brain went no further than those of its external and internal morphology. From cases reported by Hippocrates or Galen, Greek, then Arab, physicians described epilepsy, paralysis, or madness following discrete brain lesions. Battlefields provided marvelous pathological material: Napoleon was a generous source of wounded heads. Barbers were often expert at trepanation and decompressing skull fractures. Ambroise Paré was their legendary master. The nineteenth century revolutionized brain science with the birth of clinicoanatomical methodology, and progress in this field continued throughout the twentieth. Paul Broca was a pioneer of correlating a brain lesion—its localization, extent, and depth—and the victim's symptoms. There was agreement that the brain was the seat of our physical and intellectual faculties—without bothering much about the "soul," jealously preserved in the halls of the spirit. The question, "Spirit are you there?" has today lost its interest, to the benefit of algorithms. But the head has lost none of its prestige, and we need to talk "face to face" when we want to educate a man, at least according to Montaigne's precepts: "I would also wish that one take care to choose a guide who has a head well made rather than well filled."[1]

Before discussing the sick brain, we wish to dispel a major embarrassment that regularly blocks the approach to a patient. Are we dealing with neurology or psychiatry? We encounter the famous mind-body problem, which has as much to do with popular psychology as with metaphysics, and specialist practitioners are often the first to misunderstand. One of the results of May 1968 in Europe was a divorce between neurology and psychology. The former has its battalions of radiologists and biochemists, the latter its psychotherapists and its quarrelsome armadas of glib psychoanalysts. Today there is reconciliation, but the explosion of our knowledge about the brain does not seem to have solved the question of relationships between psychology and the central nervous system or, more exactly, what we think we know about the structure and function of the brain and what we think we know about mental illness. In its most trivial form, not only for a poorly informed public, the question remains: is mental illness organic or not?

In his essay on the brain Georges Lantéri-Laura talks good sense:

To get out of this *impasse*, let us first note a naïve concept which may be quite harmful. When we wonder if psychiatry refers to psychic or organic pathology we ignore the hardly challengeable evidence that our whole psychic life, conscious as well as unconscious, intellectual as well as affective, and so on, functions thanks to the central nervous system, especially the brain, and in the brain the cortex. However subtly effective reason may appear, however lively and delicate passion, they would not exist without the brain, unless you blindly support a naïve optimism that may not seem contradictory but which ignores a mass of clinical and experimental data that is hard to avoid. We can simply say that the present state of our knowledge leads us to believe that all our experience of reason, sentiment, imagination, and so on, is supported by cerebral activity, even if we do not have much concrete information on the subject. We consider ourselves refined enough to see in naïve optimism rational theorization, but we hold onto it.[2]

This problem brings us back to the soul, that intrusive stowaway in our brain. In 1908, the Swiss psychiatrist Auguste Forel said, "the living brain and the soul are one and the same."[3] Whether we choose to heal the soul or the body, we always treat wounded flesh.

Disheartening Statistics

Of the over 700 million inhabitants of Europe almost a quarter is aged over sixty-five. According to public authorities, the aging population is leading to more cases of senile and presenile dementia. As we shall see, age is not everything: the brain is sensitive to other, more damaging factors, such as boredom and affective problems. Furthermore, in addition to obvious public health implications, neurological illness poses major socioeconomic problems. According to the World Health Organization cerebral pathology accounts for 40 percent of European expenditure for public health. In Europe, we already see six million sufferers from Alzheimer disease, one and a half million from Parkinson disease, two million from the effects of stroke, and two and a half million epileptics. In all, brain disorders cost more than 800 billion euros per year.[4] Parkinson disease in the United States is estimated to affect up to a million people, with five thousand new cases each year, and it costs more than ten billion dollars per year. The figures for Alzheimer disease are even worse: in the United States some five million people have it, with up to half a million new cases each year. The annual cost is more than 200 billion dollars. The situation is hardly better concerning mental illness. In Europe one adult in four is, has been, or will be a victim in his lifetime. It represents the top cause of death in twenty-five- to thirty-five-year-olds. Tens of millions suffer from diseases such as depression or anxiety, and five million from psychosis (delirium or schizophrenia).[5] Faced with these figures, we must accept that our society remains incapable of tackling this problem other than by incarceration or the massive use of psychotropic drugs. But this has not always been the case. In ancient society madness was often seen as sacred. Killers can still be judged insane rather than guilty.

Since scientists ceased to consider the brain as a black box some fifty years ago we are beginning to understand better the nature of the disturbances responsible for neurological or mental illnesses. This rather overdue knowledge now enables us to deal more selectively with brain pathology. We shall now discuss the aging brain and the horrors of neurological or mental disease that plunge the patient into darkness, not knowing if it is purgatory or hell.

The Aging Brain

Contrary to the myth that the adult brain loses some hundred thousand neurons per day or even more if there is alcohol abuse or smoking, there is evidence that the total number of neurons does not depend as much on age as on the nature of one's neurons, some large neurons shrinking and the number of small neurons increasing with time.[6] Because the child's brain reaches 90 percent of its final size by nine or ten years of age, we long believed that thereafter its evolutive capacity was limited. However, it was discovered that even the eighty-year-old brain remains capable of structural and functional changes. On the other hand, it is certain that mental faculties decline with age if neural circuits are no longer stimulated by novelties and surprises. If we follow this principle, the saying "You can't teach an old dog new tricks" should become "If an old dog doesn't learn new tricks, it will forget its old ones."

Random sampling in a population or analysis of the same cohort over time show that not all perceptual and cognitive functions suffer the devastation of time in the same way. Certain functions decline with age while others remain stable or even improve.[7] Although operations like the mental rotation of an object in space, calculation, memorization, or reading decline with age, the opposite happens for other faculties, such as the enrichment of vocabulary. In general we consider that executive functions are the first to decline with age. Some years ago we discovered (or rediscovered) that physical reeducation or new learning contributes to slowing the negative effects of time on the brain. If his mind is not stimulated, an active person obliged to cease professional activity because of the legal age of retirement will certainly be faced sooner or later with cognitive deficits related to stopping intellectual activity suddenly. Because we live longer than we did in the past,[8] we must get used to meeting more and more persons with declining mental faculties or even dementia, for modern society no longer knows how to care for aging brains that become bored. This injustice has multiple consequences. Not only does it deprive the patient of his identity; it deprives society of knowledge and wisdom that our elders accumulated over time.

Dementia and Other Cerebral Pathology

The degeneration of neurons (neurodegeneration) is more or less rapid and often selective cell death, for example, of dopaminergic neurons in Parkinson disease. The speed of cell death depends on the processes involved. When tissue is damaged by trauma, cells die rapidly by a process called *necrosis*. On the other hand, without trauma a cell can commit suicide. This is a genetically defined program called *apoptosis*, as we already mentioned in chapter 2. Normally few of our neurons die, but when a neurodegenerative disease occurs cell death accelerates, as in Parkinson disease, where we distinguish three categories of neurons: healthy but aging neurons, neurons that die by apoptosis in a few days, and pathologically weakened neurons that die slowly over several months. To treat the evolution of this disease, should we tackle normal aging, programmed death, or slow pathological death? Many research teams have chosen to slow apoptosis. This may well be a strategic error because it only involves a tiny fraction of dying cells, at least in Parkinson disease.

It is a highly disabling disease that affects 100,000 people in France, with 8,000 new cases each year. In the United States there are said to be up to a million cases. It often begins between fifty and sixty-five years of age and manifests a triad of motor symptoms: shaking at rest, muscular rigidity, and the inability to perform certain movements. Depending on the state of advancement, patients may suffer from difficulty with elocution and even depression. Although we do not know the exact cause, it results from a selective loss of dopaminergic neurons in a brainstem nucleus, the substantia nigra (figure 1.3). Because the motor circuits are well identified, the symptoms of Parkinson disease can be treated specifically apart from trying to slow apoptosis. Administration of a substance (L-DOPA) that is converted into dopamine is one therapeutic strategy. However, when a patient receives L-DOPA, the slow movements that characterize the disease are replaced by fast movements; this is termed hyperkinesia. To avoid these complications other strategies have recently been used, such as intervening directly in the defective structures. For example electrodes connected to a pacemaker can be implanted in a deep brain structure, the *subthalamic nucleus* (figure 1.3), which is only a few millimeters in diameter. The electrodes deliver electrical stimuli to the circuits damaged by the disease. This approach allows sufferers to recover a degree of coordinated movement. Even if this technique is not yet applied widely (it is given to some 5 percent of Parkinson patients) it nevertheless represents undeniable progress. However, although this stimulation reduces or even abolishes the motor symptoms, it does not compensate for the overall dopamine depletion caused by the death of dopaminergic neurons. A major inconvenience is that it is difficult to foresee the effect on motor activity, for the subthalamic nucleus is heterogeneous, and its neurons react unpredictably.

Gene therapy is a recent alternative to stimulating electrodes. It aims to replace a defective gene and can be done using a special vector, usually a modified virus that will not replicate once in the host. Once in the brain, it can deliver the gene inside the neurons. It can also consist of modifying cells in culture before grafting them in the brain. These approaches are theoretically very promising, but we must emphasize that the technique is difficult to master because of the great diversity of neurons in the brain. Years of research are still necessary before medicine can benefit without risk from progress in gene and cell therapy for repairing the brain.

What do we know today about that other equally disabling and common pathology, Alzheimer disease? First of all, it is the most widespread neurological disease (800,000 people in France, five million in the United States) and is characterized by the irreversible loss of certain neurons leading to dementia. It is associated with the accumulation of *senile plaques* that the brain cannot eliminate. In addition, there is loss of neurons and tangles of so-called *tau protein* inside neurons. One hypothesis puts the cause of the disease as the action of tau protein associated with *neurofibrillary degeneration*, causing neuronal death.[9] Another theory, less favored by the scientific community, supposes that the disease appears when the pollutants zinc and copper accumulate in the brain. Although different theories may seem contradictory, it is possible that the disease is a result of a combination of several mechanisms.

The first structure affected is the hippocampus, which, as we have seen earlier, is part of the limbic system and named because of its seahorse-like shape, and which is indispensable for forming short-term memory, the content of which is rapidly transferred to the neocortex for long-term storage. Later the disease extends to neighboring structures and attacks those regions responsible for affect and emotion, with the appearance of episodes of agitation and confusion. Diagnosis is not easy, for at the beginning the disease can be confused with the loss of cognitive function related to aging. To establish a precise, unambiguous diagnosis two approaches are possible. Clinical examination to evaluate the level of attention can detect precursor signs. These tests depend on measurement of evoked potentials, or positron emission tomography (PET scan), or fMRI. A second, radically different approach involves genetic study. We now know some genetic determinants of the disease.[10] Among these genes, *ApoE4* is the greatest risk factor when it mutates. This gene regulates the production of a protein that facilitates the transport of cholesterol and other lipids in the blood. The presence of *ApoE4* indicates that a patient has a greater susceptibility to Alzheimer disease. Fifteen years after this first discovery, three new genes, *PICALM*, *CLU*, and *CR1*, were identified. In 2011 five other predisposing genetic factors (*ABCA7*, *MS4A*, *EPHA1*, *CD2AP*, and *CD33*) were associated with the disease. Altogether, this progress in human genetics offers several benefits. It provides deeper knowledge of pathological mechanisms underlying the disease, reinforces links between the disease and cardiovascular factors, and also identifies new risk factors associated with age. However, the presence of one of these genes is an indication, never a certainty. Only retrospective postmortem diagnosis can confirm or disconfirm the true diagnosis.[11]

We have often emphasized that the adult brain is a highly plastic organ. This inherent property is expressed throughout life, but it declines with age if the intellect is left to lie fallow. In other words, the more older people are stimulated intellectually the greater their chance of retarding the onset of disease. This being the case, a Japanese group developed *learning theory*, which combats senile dementia by using techniques based on reading aloud combined with mental arithmetic.[12] The aim is to keep the brain alert in older people by reducing sensory or attention deficits. The results seem promising. Undoubtedly environmental enrichment constitutes an effective complement to so-called classic therapy.

Diseases of the Mind

The problem of identifying the origin of mental disorders is not an easy one. We find in the literature that it is an old question. In ancient Greece, Homer's poems speak of madness as a punishment by the gods when they are offended. It is however with Hippocrates and his humoral theory that we find the first real classification of mental problems such as mania, melancholia, paranoia, and epilepsy.[13] According to him these mental pathologies could be related to four temperaments: sanguine, choleric, phlegmatic, and melancholic. During the centuries that followed this theory the causes of mental disorders remained unknown and the symptoms poorly defined, making diagnosis difficult. The twentieth century saw two rival schools of thought: one defended the psychodynamic or psychoanalytic hypothesis (itself divided into different factions); the other was organic and localizational, trying to explain and treat mental illness in the light of neurobiological, psychopharmacological, and genetic discoveries.[14] Today, treatment of mental disorders attempts to reconcile these two tendencies: biological treatment, usually with medication but not always (see the next chapter), is accompanied with psychotherapy.

Neuropsychiatrists throughout the world collaborate to standardize diagnostic criteria of these disorders in a reference work, the *Diagnostic and Statistical Manual of Mental Disorders,* of which the fifth edition (DSM-5) was published in 2013. Fifteen years after its previous revision this manual remains a standard for consensus among psychiatrists. Recent discoveries form the basis for this revision. Its objective is to give a new dimension to psychiatric diagnosis. Until now the aim was to establish precise categories with clearly defined properties and sharp borderlines, but much mental pathology resisted such categorization. So psychiatrists chose a multidimensional approach to establish scales to define clinically homogenous categories. Their hope is to identify more easily the physiopathological mechanisms triggering psychiatric disease, for example, taking into account the number of episodes or the nature of the latest one (such as depressive, hypermanic, or hypomanic), thus permitting the definition of different clinical forms of, for example, depression.

The arrival of molecular biology and behavioral genetics has permitted great advances in mental health. Genes involved in the transmission of susceptibility to depression, schizophrenia, and autism have now been identified. Family studies on twins and adopted children demonstrate the determining role of genetic factors in the transmission of certain psychiatric diseases. For example, in autism the risk of both identical twins being affected is five to ten times greater than in nonidentical twins, and the risk of being affected oneself when a sibling is affected is ten to fifty times greater than in the general population. The future therapeutic arsenal for the treatment of psychiatric illness should evolve toward therapy of the individual. To provide a patient with optimal medication means analyzing his genetic background: such should be the offer of today's personalized medicine to tomorrow's psychiatry. Aldous Huxley or George Orwell would perhaps not have disapproved of such scientific progress.

When the Brain Takes Over a Lost Limb

The anatomical organization of the sensory or motor cortex is described in terms of areas. Sensory organs project to specific primary cortical areas. Precise maps of the body have been identified by moving a recording electrode over the cortex while stimulating the skin. The whole body surface is represented across the cortex, but this representation is not faithful to physical reality. There is a deformed projection of our body in the cortex such that our ability for movement and perception are based on coded representations in our brain, the whole forming our body image. Like shadow puppets, these cortical maps form miniature human shapes in our brain, the sensory and motor homunculi that we discussed in chapter 3. The relative size of sensory representations in the cortex varies according to modalities in different species: the vibrissae are dominant in the mouse, the tail and face in the monkey, and the hands and mouth in man. Although genetically determined, the contours of these maps differ from one subject to another in the same species, for they are constantly redrawn by sensory and motor experience.

We owe the first demonstrations of major reorganization of the maps following sensory deprivation to two American researchers, Michael Merzenich and Jon Kaas in the 1980s. Since then, many observations after behavioral manipulations or accidents have confirmed the ability of the cortex to redraw its own maps. The representation of the hand shrinks if the fingers are immobilized or its peripheral nerves sectioned. On the contrary, if a primate is trained to use only the middle finger of its hand, the area representing this finger encroaches on that of the other unused fingers. This is one more example of the obsolescence of the old debate about the innate and the acquired and about how plastic the central nervous system really is. These data emphasize the error made by Robert Schumann who, to perfect his pianistic technique, is said to have worn an apparatus to immobilize a finger of his right hand during his exercises. The result was disastrous: a year later he ended his career as a virtuoso because of growing difficulty to move his fingers with precision.

The extreme plasticity of cortical maps is also evident in *phantom limbs*. This phenomenon is found frequently in patients who have a hand, arm, or leg amputated, and it manifests as a sensation that the lost limb is still there. It occurs in about 95 percent of amputees, and some 70 percent suffer pain from the limb, at least during the first weeks after the accident or operation. The sensation was first described in the sixteenth century by Ambroise Paré, surgeon to Charles IX, but the term was first used by the American Silas Weir Mitchell in 1871. However, even if the phenomenon was known in medical literature since the nineteenth century only recently was the mystery elucidated. The phantom pain occurs when the nerves that normally innervate a limb lose their cortical target. Vilayanur Ramachandran at the University of California, San Diego, explained the phenomenon. He wondered why someone who had lost his hand in an accident still felt it. He proposed that a phantom limb resulted from irreversible interruption of sensory pathways taking information from hand to cortex, and cortical zones near the hand map (arm and face for example) occupied the vacant space. Phantom limb phenomena were attributable to the reorganization of the homunculus. To prove his theory Ramachandran and his colleagues showed that by stroking the face the patient felt sensation in the lost hand. Using magnetoencephalography (MEG) to visualize cerebral activity, they demonstrated that the somatosensory cortex had undergone major changes. To treat pain from phantom limbs Ramachandran developed an approach based on a mirror box, in which the patient saw the reflection of his good arm in place of the lost one, which permitted artificial visual feedback. Phantom pain might be because of incoherence between the brain's intention to move the limb and the absence of response from it. The pain could be the consequence of a reaction by neighboring territories trying to compensate for this lethargy. Use of the mirror allowed the silenced maps to work again, as the mirror created an illusion of movement in the lost arm. The patient was able to "move" the phantom limb and so reconfigure the modified sensory map. It was necessary to convince the brain that the lost limb was still there. This therapy is helping to dissipate chronic pain, a major advance thanks to the amazing plasticity of the brain. It may provide long-term benefit, although results are as yet uncertain. Other neurologists have chosen, with some success, to immerse their patients in virtual reality, giving them an avatar who has all his limbs (see the next chapter). A few sessions of imagining oneself in this virtual world is enough to trick the brain and dispel the phantom pain.

This marvelous capacity of our brain to remodel its neural networks and so reprogram its activity after a lesion or loss of a limb is astonishing. It also works when a patient receives a limb graft. At the Institute of Cognitive Science in Lyon, Angela Sirigu showed how a recipient's brain integrates a grafted limb. Using transcranial magnetic stimulation and evoked potentials she observed that the motor cortex corresponding to a hand grafted some years earlier could recover from the shrinkage noted before the graft.[15] When a patient loses a hand, the brain adapts to the modified body, deleting the motor and sensory representations of the missing limb. To do this, the neurons innervating the lost hand muscles progressively invade the arm muscles. This redistribution is accompanied by *cortical exclusion*, the disappearance of the representation of the limb in the brain. The process was for long thought to be irreversible after a certain period. But contrary to this dogma the neurologists from Lyon found with surprise that after a hand graft the motor and sensory cortex that had abandoned a severed hand could reappropriate the new grafted hand.

The Wounded Brain, the Incomplete Brain

Many invertebrates can regenerate a limb or an organ, including the nervous system. So-called higher animals, especially man, seemed to have lost this potential. Nevertheless they possess a certain facility for repair that can overcome some handicaps, at least partially. An example of this capacity for adaptation is the possibility for a function, one lost after a lesion in a cortical area, to be taken over by the homologous region in the other hemisphere. Harvey Levin of Baylor College of Medicine in Houston studied an adolescent who had no visuospatial difficulties in spite of a childhood right parietal lesion. On the other hand, he encountered immense difficulties with calculation. Clinical examination revealed that the left parietal lobe had replaced the right, interfering with its primary function, calculation. Another equally spectacular example is furnished by thirty-seven-year-old M. M., who had only half a functional brain since birth. In spite of this major handicap this young woman received normal schooling and obtained her high school diploma without difficulty. According to her teachers she had an exceptional ability in retaining figures. When they examined her brain with MRI, her neurologists discovered that a prenatal lesion had removed the left part of her brain. Remarkably, the right hemisphere was able to replace the functions normally provided by the left. The possibility for certain neural circuits to adapt to take charge of a new function was already known in the case of people blind from birth reading Braille. In them, touch activates occipital cortex intended for vision. So, areas adjacent to the lesion "recycle" their unused territory by *map expansion*.

Another example from vision concerns *blindsight*. Some people are blind but still possess what is called blindsight. They are able to move around and avoid objects that they cannot see consciously. One patient suffered a vascular accident in the left primary visual cortex in 2003, then two months later another on the right. Although the retina and the visual pathways remained intact after the two lesions, the patient was blind because the primary visual cortex did not respond to signals. Nevertheless when shown a portrait of someone expressing fright, he showed signs of fear, although claiming not to see the face. Surprisingly subjects with blindsight feel emotions expressed by a face or a posture without being able to identify the object with which they are confronted. Such clinical cases illustrate the close relationship between conscious and unconscious vision. In the case of blindsight subcortical and cortical regions outside the primary visual cortex ensure the residual vision after some compensatory reorganization. The superior colliculus of the midbrain (figure 1.2) seems to play a central role in this reorganization. It receives information from the retina and relays it to the secondary visual cortex. In blindsight involving emotion, another structure, the amygdala (see chapter 3), participates in the unconscious experience. This all shows that the brain of a patient, even an adult, can be innovative after a lesion or a malformation, to compensate for a lost function, which is taken over by neighboring circuits.

The Brain Under Stress

When we experience an unpleasant, potentially dangerous event, our brain chemistry prepares us for flight, an imminent fight, or to inhibit our actions and opt for the status quo. Of course the chosen behavioral strategy will correspond to the situation that provoked this stress, but it will also draw from past experience. Adrenaline is the hormone to save the situation: it enables us to mobilize vital energy for flight or fight. It is the physiological response to acute stress, which is experienced as acceleration of heart rhythm, raised blood pressure, palpitations, and shortness of breath. We feel these effects when, for example, we are walking in the yard and notice in the distance something that looks like a snake but is only a garden hose. Happily, this symptomatology lasts no longer than a matter of minutes. The phylogenetic origin of this form of quick emotion is very distant. It appeared with the limbic system of mammals. This acute reaction to stress can be exciting and even beneficial in alerting our senses.

On the other hand, when such events are repeated indefinitely too much stress ends up being exhausting or even a danger to health. This chronic stress represents a situation that progressively occupies the whole body and mind and results in the subject seeing no favorable issue. Hopeless, he gives up looking for a solution and accepts the unpleasant situation. High levels of glucocorticoids can be detected and remain in the system for years.[16] As we have seen, the hippocampus plays an important role in learning and declarative memory (memory of what we can describe verbally, rather than skill). It is very exposed to the dangerous effects of high doses of cortisol. Consequently, chronic stress can cause deterioration in memory, which disturbs learning and adaptation. The other harmful effect of high cortisol, and thus of stress, in the hippocampus is an increased susceptibility to depression.

So although we are relatively well adapted to physiological effects of acute stress (such as ancient man experienced when seeing a predator), we remain vulnerable in situations that lead to chronic stress. We now understand the mechanism by which chronic stress alters serotonin receptors (increased 5-HT2A and decreased hippocampal 5-HT1A). Similar modifications in receptors are seen in suicide victims, and prolonged administration of antidepressants produces opposite changes to those of chronic stress.

We have known for some time that depressives have a hyperactive hypophyseal-pituitary-adrenal axis (HPA) and that prolonged activation of the HPA tends to favor depression. Chronic stress, by challenging the HPA, leads to basically irreversible structural changes in certain brain regions. In particular, the hippocampus suffers considerable loss of neurons during prolonged stress, notably related to the desensitization of glucocorticoid receptors, leading to negative trophic effects. In sufferers from diseases that overproduce cortisol (for example Cushing syndrome) we see a high incidence of depression that becomes treatable when the cortisol level reaches normal levels. The final product of the HPA, glucocorticoids, plays a role in depression by influencing several neurotransmitter systems, including serotonin, noradrenaline, and dopamine, the chemical triad strongly implicated in depression.

Cognitive Remediation

Cognitive remediation is a process of training and learning aimed at cortical areas involved in attention, learning, memory, planning, and execution. The method uses techniques conceived to improve cognitive performance in patients whose mental functions have been altered, for example by stroke, tumor, trauma, early Alzheimer disease, depression, or schizophrenia.[17] The objective of cognitive remediation is to support weakened cognitive capacities that are still present and then learn new compensatory strategies. This neuropsychological technique seeks to encourage patients to resolve problems and use solutions to improve cognitive performance (such as memory, attention, and reasoning), thus favoring better cognitive autonomy.

Therapies like cognitive remediation, or cognitive behavioral therapy, and other programs are more and more part of the therapeutic arsenal that seeks to offer patients progressively more autonomy in cognitive performance. The mechanism of how these therapies improve mental faculties is not always clear, although we are beginning to realize that they concern not only changes in specific aspects of brain function (directly on plasticity of neural networks or indirectly via cerebral vasculature) but also more general changes at the behavioral level. In any case, whatever the underlying mechanisms, cognitive remediation depends on adult brain plasticity.[18] William Spaulding and his colleagues evaluated change in schizophrenics after cognitive rehabilitation.[19] They identified improvements in nine of twelve measures of cognitive function, including memorization, visual masking, and tasks involving executive functions. In general, changes were more robust for high-level cognition, essential everyday functions like planning activity, correcting errors, or seeking new solutions. By reorganizing cerebral activity and exploiting the undervalued resources of brain plasticity, cognitive training not only improves the patient's cognitive function but equally helps social and professional reinsertion, thanks to better behavioral adaptation to his environment.

Sometimes cognitive remediation can be assisted by computerized virtual stimulation of mental faculties needed for everyday activity. The pioneering work of Michael Merzenich in San Francisco demonstrated the importance of training in phonetic processes, considered to be severely altered in dyslexic children, to improve reading skills.[20] Since then this approach has had some success. Merzenich was surprised to discover that children suffering from language troubles saw their performance improve after daily training with a computer game. A program "FastForWord" has been developed specifically to remedy the deficit in auditory processing that Merzenich considered a cause of learning problems.[21] This work opened a new field of research that is proving more fertile every year: computer-assisted neurodevelopmental remediation. Several programs have proved the importance of this therapeutic approach in the treatment of hyperactivity and attention deficit in dyspraxia and dyslexia.[22] Computer-assisted cognitive remediation also seems to be a valuable interventional tool for the treatment of Alzheimer disease and has a measurable impact on cognitive deficits related to dementia. For this there exists cognitive remediation software designed to stimulate attention, memory, language, motor coordination, and visuospatial skills in patients. In all these cases multimedia training by computer seems to stimulate neuronal plasticity, and computing activity facilitates the acquisition of compensatory strategies for the adult brain.

Compensation, stimulation, even repair—but our path does not halt here. Let us look at new possibilities for modification and optimization opened by new brain sciences.

6

THE ENHANCED BRAIN

To desire immortality is to desire the eternal perpetuation
of a great mistake.
—ARTHUR SCHOPENHAUER

CAN WE IMAGINE A ROBOT that can read our thoughts and carry
out unpleasant daily tasks at our convenience? This subjugation of a
robot by our thoughts would constitute a victory of man over machine.[1]
Or is it only fantasy, a derisory product of science fiction? No, it is a
dream that could become reality, given that scientists can record brain
activity accompanying the planning and execution of a voluntary movement and replay the tune to a machine to execute the task. So, after suc-
ceeding in transforming matter into energy, then energy into work or
information, humanity is embarking on a new transformation: thought
into action! We shall once again see the brain adapting itself to the most
incongruous situations.

Recruited from a variety of disciplines, technoscientists are indeed
at work to short-circuit thought directly to action, independent of the
peripheral nervous system. The first fruits of this research are already
ripe: sometimes fascinating, sometimes frightening depending on the
circumstances. Henceforth the capacity of thought to act at a distance is
not magic on a par with spoon benders and chain breakers. There is no
need for complex equipment such as imagined by directors of science-
fiction movies. The technology needed for thought to control a machine
at a distance is relatively simple. It suffices to don a helmet to record our
mental activity in the form of an electroencephalogram (EEG), already

mentioned in chapter 4, and transmit this information to a computer, which sends commands to servomotors. Signals recorded from the cortical surface are derived from neuronal electrical activity. There are two ways to record them, one invasively, the other not. Noninvasive techniques involve electrodes on the scalp that record a large number of neurons to produce an EEG. Spatial resolution is relatively low, and it is difficult to localize precisely the origin of the signal, and the duration of recording is limited to a few days. In invasive recording an electrode or grid of electrodes is implanted through the skull. This technique offers excellent spatial resolution because it allows the measurement of the sensitivity of a neuron or micronetwork of neurons. A grid of electrodes on or under the dura records a precise local *electrocorticogram*. Cerebral electrical activity is more or less complex to process depending on whether it is from one neuron or from thousands in the form of cortical slow waves or evoked potentials.

With an EEG helmet we can now control movements of an artificial limb simply by thought, with no muscular effort. This technology, the *brain-machine interface* (BMI), aims to establish a dialogue between human thought and machine function, whether a computer or a robot.[2] This field of research, at the crossroads of the fundamental, experimental, and clinical, is in full swing. In 1994, at the time of the first interfaces, only half a dozen laboratories were interested in this technology. Five years later around forty were represented at the first international conference dedicated to these machines. In 2007 there were nearly a hundred participants,[3] and that figure had doubled by 2010. A few examples will help to measure its extent.

Born of the fusion of disciplines including engineering, neuroscience, informatics, mathematics, physics, and medicine, brain-machines will radically modify the way we interact with our environment, whether we are healthy or not. Quite a defeat for fanatics of the soul and its mysteries! Unless it is simply a defeat for man, according to the lamentations of confirmed humanists. Just imagine: the development of these interfaces is the promise of a future world where we shall move around, use our computers, and communicate merely by thought. "Think and the machine will do it for you" will be the popular adage. Several prototypes are already at work throughout the world, especially in the United States, northern Europe, and Japan. These first robots reflect the progress over the last fifty years. They also reveal how much remains to accomplish so that one day the promises of technoscience may become part of the therapeutic arsenal of neurosurgeons and other practitioners seeking new tools to alleviate brain disorders. New perspectives for rehabilitation of motor and sensory deficits, especially of sight and hearing, are already tangible. Announced with special effects borrowed from science fiction, brain-machines are progressively leaving the world of film and fiction to invade the real world of service to the person, whether in good health (the enhanced brain) or to assist in overcoming handicap (the rehabilitated brain).

The Cerebots Are Coming

A BMI is an apparatus for communication at a distance between a brain and a machine. For this apparatus to function harmoniously, communication must be reciprocal so that information can pass first from brain to machine then from machine to brain.[4] Information from the environment must be gathered (usually through sensory organs but also from mental representations formed from the sensation) and then action taken on this information (for example, lifting a glass or kicking a ball). Considering these two functions related to perception and action,[5] the fundamental principle of these interfaces is to link thought to action without muscular activity. The machine involved could be a computer, a robot, or a hybrid that we shall call a "cerebot."[6] To be effective, the BMI must carry input and output channels. The input allows acquisition, amplification, and digital processing, via a series of algorithms, of signals from the brain performing some cognitive task. The output analyzes the signal from the brain to define a series of orders to control the machine. In order for the inputs and outputs to work harmoniously, a feedback loop provides for learning by a continuous series of readjustments through trial and error. We shall come back to the notion of this "biofeedback," which is central to the concept of utilization and optimization of human BMIs.

Thanks to this hybrid machine, an individual can communicate with his more or less close environment without need to use nerves and muscles.[7]

As often in science, this revolutionary idea is not only the fruit of recent progress. The principle is based on some fortuitous research begun more than half a century ago but not rigorously controlled. The first success with BMIs dates from 1957, with the fitting of the first cochlear implants.[8] Rather than amplifying sound with a hearing aid, these experiments stimulated the cochlea of the inner ear by surgically implanted electrodes to transmit signals via the auditory nerve to the brain. Initially the perceived sensation was not like normal perception, nor like that from a hearing aid. But after a relatively long period of learning, with the help of a speech therapist, patients learned to retrain their hearing to make sense of direct stimulation of the cochlea, that is, to construct a percept. There was rapid clinical success, even if the procedure still needs improvement, as we shall see later. Today, it is estimated that about 100,000 people throughout the world have cochlear implants. Around a third of implanted children obtain excellent results, with comprehension equivalent to that of children with intact hearing; another third are able to understand speech; and the last third have difficulties, sometimes because of the implantation being carried out too late.

Hearing is not the only sensory modality to benefit from "high-tech" methods and bionic organs (neuroprostheses). Vision can equally be rehabilitated with an invasive electronic apparatus. Artificial retinas are being produced to compensate loss of photoreceptors leading to blindness in such pathology as retinitis pigmentosa or age-related macular degeneration.[9] Even in severe cases of these disorders, some neurons of the retina remain in contact with higher visual centers. These contacts are functional: clinical studies have shown that electrical stimulation of residual neurons produces visual images. Research on retinal prostheses was the subject of a pilot project by the U.S. Department of Energy. The apparatus consisted of spectacles equipped with video cameras to capture images. They were connected to a miniature microprocessor, on the patient's belt, which interpreted the video images and transmitted the information to the retinal prosthesis, a stimulator with several electrodes. The retinal photoreceptors were thus replaced by an electronic system stimulating residual retinal neurons to produce images in the visual cortex.

In 2002, the University of Southern California and Second Sight Medical Products, leaders in the field of retinal prostheses, announced that volunteers had tested an artificial retina named Argus I, which had only sixteen electrodes and therefore low resolution. Nevertheless, between 2002 and 2004 it enabled six people with retinal degeneration to recover rudimentary vision. They could detect a door or window, avoid large obstacles, and even read characters about thirty centimeters in size. These first achievements allowed the concept to be validated and for a large international research program to be launched to improve the performance of artificial retinas. Then a second generation of retinal implants called Argus II, with sixty electrodes, was developed. The prototype was tried in 2007 on volunteers, first in Mexico, then in the United States, and finally in Europe in three centers: the Ophthalmology Service at the University Hospital in Geneva, the Quinze-Vingts Ophthalmology Hospital in Paris, and Moorfields Eye Hospital in London. In Europe two separate clinical trials of this new generation of prostheses were conducted in 2008. They evaluated the impact on vision of a medium-resolution retinal implant developed by Mark Humayun at the Doheny Retinal Institute of the University of Southern California.[10] In 2010, ophthalmologists in Geneva, in collaboration with colleagues in Los Angeles and Paris, succeeded in implanting a higher-resolution implant in a totally blind patient.

We should note that all these operations aim to recover a certain degree of vision in people suffering from hereditary blindness such as retinitis pigmentosa. Although these early operations achieved a certain success clinically, researchers' objectives were limited to a demonstration of reliability of the method. First results allow an evaluation of the impact of the method, but their main importance is to solicit improvements. Progress is encouraging, and ophthalmologists confirm that patients can perceive light, forms, and movement; they can move about following a line on the floor and can distinguish a person at six meters. But results vary between subjects. Although some patients can read small text on a computer, others remain unable to use the visual information sent to their cortex. However, these trials show that it is possible to reactivate the retina. A third generation, with resolution improved to two hundred electrodes, is being developed. But, hungry for novelty, researchers are already at work on artificial retinas consisting of more than a thousand electrodes. Eberhart Zrenner at the Institute for Ophthalmic Research in Tübingen has produced an implant with 1,500 electrodes.

Thomas Elbert's team at the University of Constance demonstrated the amazing ability of the human sensory cortex to reconfigure itself according to circumstances. As we saw in chapter 3, sensory information from the body surface is transmitted to the somatosensory cortex. For example, when we touch an object with a finger signals along sensory nerves trigger a response in the corresponding part of the cortex. Using noninvasive MRI techniques, Thomas Elbert and his colleagues studied somatosensory cortical activity. They wished to test whether the adult brain could undergo major reorganization with important morphological and functional consequences. This plasticity was observed in blind people reading Braille (see chapter 4).[11] There were surprises when comparing cerebral images obtained in the blind reading Braille with three fingers, those using only one, and sighted subjects not reading Braille. The researchers noted that the region of the cortex devoted to the hand was much larger in the three-finger readers than in any other subjects. In other words, the brain was able to reconfigure itself depending on the Braille method chosen. This result was easily confirmed using a tactile test: three-finger readers found it hard to say which finger was touched when stroked with a feather. This experiment showed that the brain was able to fuse information from each of the three fingers to produce a single, more robust and reliable message, which is so important for the Braille reader. It is obvious that the representation of different parts of the body in the somatosensory cortex can change rapidly to adapt to the needs of the moment and personal history. Thanks to this basic property, the adult brain can construct mental representations from electrical impulses delivered by artificial sensory receptors.

With the rather successful arrival of brain-machines in the field of sensation, clinicians turned their attention to the repair of motor function. This was facilitated by fundamental information acquired more than forty years ago. For example, Edward Evarts and his colleagues of the National Institutes of Health (NIH) at Bethesda, Maryland, in 1966 recorded electrical activity in that part of the cortex of the macaque responsible for planning voluntary movements and deciphered the information contained in it.

I Think, Therefore I Act

Uses for brain-machines already figure in several aspects of our life.[12] They penetrate important sectors of our society, like health, communication, and security. In the field of health, it is mainly in the rehabilitation of a handicap, by prostheses or text spellers, for example.[13] According to the United Nations, some 500 million people throughout the world, of whom some 80 percent live in developed countries, suffer from a serious mental, physical, or sensory handicap. In communications, the numerous uses of brain-machines concern mainly the field of multimedia (games, virtual reality, telephone, and Internet).[14] Finally, remotely controlled robots for bomb disposal or other military use are examples of security-related interfaces.[15]

We should be quite clear. Medicine remains one of the major applications of BMIs. Numerous disorders are targeted by these new technological possibilities with, at the front line, the extremely debilitating pathologies that cause major paralysis. Some head traumas and brainstem vascular accidents provoke total loss of voluntary movement, apart from some muscles in the lips or eyelids. In such cases, patients are perfectly conscious of their state but unable to communicate with those around them. This is the *locked-in* syndrome, with intact consciousness but an imprisoned, immobile body. The painful experience of the imprisonment of a lucid mind in a body in a lead coffin is described in the memoirs of a victim of a brainstem lesion. Imprisoned in his "diving bell," Jean-Dominique Bauby was hospitalized at the age of forty-four. His mental state was intact, and he was still able to blink. By using the only means of communication he still had, moving his left eyelid, he described the existence of a world in which nothing remained except completely disembodied thought.[16] In 1998, Philip Kennedy and Roy Bakay performed a cerebral implant in a patient affected by this syndrome, which enabled him to position a mouse cursor using his thoughts.[17] By moving it over the virtual keyboard he learned to spell words by imagining them. Thanks to this BMI, contact with the outside world was reestablished.

These interfaces are also a source of hope to minimize the motor problems of partial paralysis like those following spinal cord traumas in which the victim loses use of a limb. In 2005, a patient suffering from amyotrophic lateral sclerosis moved an artificial arm using only thought, thanks to the implantation of a grid of electrodes directly in his motor cortex. Miguel Nicolelis of the Duke Center for Neuroengineering in Durham, North Carolina, was a pioneer, one of the first to demonstrate that it was possible to control the movement of an artificial arm merely by reading the electrical activity of a specific cortical area. In 2003, his team published astonishing results from their first BMIs.[18] A monkey was trained to manipulate a joystick in a computer game to displace a spot on a screen. When the monkey succeeded in touching a circle with the spot, it received a reward of apple juice. Microelectrodes were implanted in the motor cortex, which is involved in voluntary movement (see chapter 3). The signal emitted by the neurons controlling motor activity was used simultaneously to move a robot arm placed at the monkey's side. When the joystick was disconnected from the computer, in order to continue to play and earn the reward, the monkey had to imagine the movement necessary to displace the robot arm. Gradually, by observing the spot move across the screen, it learned that it could move it without physically intervening on the joystick. For the first time, the conceptual barrier of action without body movement was removed.[19] The monkey had just learned to move a robot arm simply by thought. For this exploit, *Time* magazine placed Miguel Nicolelis on its list of the most influential scientists of 2004.

What is the basis of this learning that allows the transmission of a motor command at a distance? Have the neurons of the motor cortex been reprogrammed by learning to take control of the robot arm? Or have they acquired a new function while still retaining their old motor program so as to control both the biological and the robot arm? The second hypothesis seems the more likely because the use of the extra artificial arm in no way diminishes the monkey's ability to use its natural arm. The Nicolelis team even showed that the monkey could use its three arms simultaneously to undertake different tasks.

In view of progress in this particular field of technoscience, in 2006 the Pentagon, through the intermediary of the Defense Advanced Research Projects Agency (DARPA), invested fifty million dollars in a program called Revolutionizing Prosthetics to develop an artificial articulated bionic arm. The aim? To construct an artificial limb from shoulder to wrist, ending with a hand equipped with an opposable thumb and mobile fingers. Four years later, the Rehabilitation Institute of Chicago (RIC),[20] backed by a group of private enterprises, succeeded in inventing a bionic arm meeting those specifications. The same institute had previously acquired a high reputation for rehabilitation by permitting Jesse Sullivan, a fifty-five-year-old electrician who had lost both arms by electrocution, to recover the use of his upper limbs. With the latest version of this robotized prosthesis, which was developed by Todd Kuiken of RIC, Sullivan was again able to lace up his shoes, tie his necktie, shave, and lead an almost normal life.

Following in the footsteps of Miguel Nicolelis, Todd Kuiken then decided to use the nervous pathways that control the arm and hand to capture motor commands from the brain. Claudia Mitchell, a twenty-five-year-old woman who had just lost an arm in a motorcycle accident, was operated on to graft her sectioned nerves to a pectoral muscle. So when she planned a movement, the innervated muscle received nervous commands and produced signals captured by electrodes on the skin over it. It then only remained to send these signals to the prosthesis, which immediately translated them into motor commands. Thanks to sensors and servomechanisms in the artificial hand, the prehensile force could be adapted to the fragility of the object grasped. So, bionic limbs directly controlled by thought are ready to quit the closed universe of the laboratory and radically change the quality of life of millions of victims of accidents or pathology that deprived them of the use of one or more limbs.

All results so far indicate that the repeated use of nervous pathways during learning reconfigures those pathways significantly. This functional plasticity allows the use of an artificial arm as if it belonged to one's own body. This new body image appears and is refined through training, thanks to visual feedback (biofeedback), like the screen on which the monkey mentioned above followed the movement of the robot arm. This visuomotor loop is simply the recapitulation of a mechanism of assimilation that is at work in the monkey and in human subjects every time we try to adjust how we use a tool to increase its precision. This skill in manipulating all sorts of instruments, from professional tennis players to goldsmiths, shows that we are able to adopt a tool so that after long practice it is perceived as an extension of our own body.[21]

It is mainly thanks to this functional plasticity of the cortex that man can adapt to a brain-machine and move objects by thought. To be really effective, new research must aim to increase cerebral plasticity. To do this it must multiply the sources of information for the BMI to facilitate the reconfiguration of neural circuits. Most current experiments are based on plasticity triggered by a single source, visual feedback. In order to command very precise movements, such as grasping a fragile object with an artificial hand, other feedback loops must be found. Sensors could be installed on the limb to signal position, texture, and prehensile force. Such sensors would then flood the brain with important information about the environment and the dynamic properties of the limb.[22]

Other important progress to establish communication between a subject and his environment should be mentioned, such as portable computers with virtual keyboards in the form of a grid of letters of the alphabet. The subject, with his interface, focuses his attention on certain letters on the grid. The computer detects the responses of the subject's brain to these visual stimuli to determine congruent letters chosen previously. Using this equipment, Peter Brunner of the Wadsworth Research Center in New York wrote the word "BONJOUR" using only thought. To perform this trick before amazed onlookers, he wore a helmet equipped with about twenty electrodes, which captured and digitized the electrical activity of the cortex during thinking. The information was transmitted as an EEG to the computer, which identified the brain waves that signified execution of the planned instructions by the cortex. How was this trick performed? When Brunner concentrated his thoughts on the letter B, he fixated the screen on which rows of letters and symbols appeared randomly. As soon as a vertical or horizontal row contained a B, his brain reacted to that planned signal by emitting a distinctive waveform. In fifteen seconds (or less with training) the computer determined the chosen letter. It had learned to detect a congruence between what the EEG recorded during the subject's exposure to a visual stimulus and the letters on the screen. You can imagine the possible applications of this technology to patients with language disorders.

Niels Birbaumer of the University of Tübingen, whose specialty is to help patients with amyotrophic lateral sclerosis, performed another exploit. At an advanced stage of the disease these patients can no longer move. In the 1990s, using fMRI to detect different states of conscious-ness, Birbaumer and his team discovered that the greatest distress suf-fered by their patients was not immobility but depression induced by their incapacity to communicate. The team then sought a procedure to reestablish contact between patients and their families. When patients began to use an interface to decode EEG records to control a computer, their condition improved as they emerged from their isolation.

Thanks to BMIs, patients suffering from chronic pain can regulate their pain perception by activating the brain region concerned with pain, which can be identified by imaging. Sufferers from Parkinson disease can considerably reduce their movement disorders by applying deep-brain stimulation to their basal ganglia. This technique was developed in the monkey by Bernard Bioulac in Bordeaux, and in 1993 the team of Alim-Louis Benabid in Grenoble succeeded in controlling the motor symp-toms of Parkinson disease by stimulating electrically specific regions of the brain. In 2009, Miguel Nicolelis went further by delivering electrical stimulation to the spinal cord rather than the brain. His first experiments on animals were successful.[23] In a few seconds, stimulation of the spinal cord improved mobility in rats and mice with an induced Parkinson-like state. The rodents' motor activity was improved thirtyfold, and their movements were more rapid and precise when light stimulation was de-livered. The next stage of this research will be to use the same strategy in nonhuman primates, before progressing to human patients.

We might also mention recent work that shows that BMIs can be use-ful in treating certain psychiatric disorders. The most eloquent example concerns children with attention-deficit hyperactivity disorder (ADHD). This is a behavioral syndrome with three components: inattention, hy-peractivity, and impulsiveness. Obviously these are characteristics of all human beings, but ADHD is diagnosed when they are pronounced and prolonged in children.[24] It is impossible to predict ADHD, and its causes remain uncertain. Nor is there a curative treatment. However, a study showed that children with ADHD improved their cognitive performance significantly after intensive training to act directly on their own brain activity.[25] This *neurofeedback* permits improvement in all of the primary symptoms of ADHD. Sensors are placed on the child's head and con-nected to a screen, which displays EEG signals.[26] This equipment offers the child the opportunity to observe his own state of attention during a specific task. He can therefore practice reproducing, maintaining, or reinforcing it. Neurofeedback allows for access to a different level of our consciousness. In this way, the ability of the brain to correct itself can be fully realized. This capacity for autoadjustment is an essential part of the normal function of the brain, as we saw earlier. The repeated practice of cognitive functions could improve durably the capabilities of the subject, by a process not unlike a physical-training program. The effectiveness of these cognitive-reeducation programs is now being explored in adult ADHD. By comparing treated and control groups an Australian team has demonstrated the effectiveness of a series of eight sessions of reeducation aimed at attention, planning, executive control, control of anger, and self-esteem.[27] As Jean-Michel Guilé in Montreal showed, it is now possible to broaden the scope of treatment of childhood ADHD using computerized tools for cognitive remediation similar to those we saw in the previous chapter, tools that now form part of the nonpharmacological treatment of ADHD.[28] Comparison of results by these methods and those by effective medication, such as Ritalin (methylphenidate), emphasize their equivalence or sometimes even the superiority of neurofeedback over drug therapy.[29] Thanks to this success clinical trials are under way to measure the impact of neurofeedback on other mood disorders, such as depression.

These few examples demonstrate the diversity of applications of BMIs in the field of neurological and psychiatric disorders. Indeed, whether we are discussing mind or matter, it is always the brain that is concerned. They also show the extent to which this high technology is invading our daily life. It progresses ever more rapidly, with improvements in recording and processing of cerebral activity. In the end, tens of millions of patients throughout the world, including sixteen million with cerebral palsy and at least five million with a spinal cord lesion, might benefit from the perfection of brain-machines and regain some degree of autonomy. In France, this would affect 157,000 patients with head trauma, of which there are 12,000 new serious cases per year and about a thousand spinal cord lesions per year—and this is not including the large number of people suffering from psychological problems. All these examples show how much the emergence of these interfaces, although feared by some, is welcomed by others.

The Neurologist's Rosetta Stone

To be effective, the interface must contain a small number of components functioning harmoniously. It must have an input system to acquire, process, and decode brain signals and an output to the environment, such as a virtual keyboard, a wheelchair, or an articulated prosthesis. Finally, a feedback loop is needed so that the user can master the interface and, reciprocally, so that the interface can refine its interpretation of the patient's brain waves. We invite you to follow us through the pathways of the brain where thought is born from electricity.

To be decoded by a BMI the brain signals must fulfill two criteria: easy triggering by a mental task and easy distinction of signal from noise. Calculation or the mental rotation of an object in space are among the easiest activities to use with a BMI, for they fulfill these criteria. Stanislas Dehaene of the Collège de France showed how brain activity translates information perceived during calculation or rotation.[30] This could help evaluate the chances of comatose patients recovering and the likelihood of sequelae, to give only one striking example.

The therapeutic use of BMI is long and tedious because it requires successive learning efforts to match the subject to the machine. Invasive interfaces have been developed and applied in monkeys, the only animal with almost human dexterity, and are already in use. We have seen that it is possible, using a robot finger activated by artificial muscles, to use brain signals from the monkey to replicate its finger movements. By reconstructing the movements from prerecorded signals in the monkey it is possible to extend this motor control to several fingers and attain sufficient dexterity to manipulate objects. There is no doubt that such technological advances will radically change the quality of life of elderly people or those with sensory or motor handicaps. It is probable that these advances will penetrate all human activity, perhaps faster than specialists predict. Of course, as in the past, this progress raises fears and social, cultural, and moral questions, sometimes well-founded ones. We must remember, however, that BMIs were born of progress in techniques for treating physical or mental handicaps that were untreatable otherwise. Should we complain? Can humanity turn its back on such technology? Two opposing views emerge. If the interface concerns pathology for which a biological/electronic interaction is indispensable, ethical objections fall silent when confronted with possible improvements to the quality of the patient's life. We easily accept the idea, for example, of people with a cardiac stimulator or pacemaker. On the other hand, concerns emerge about fusion of man and machine. There is widespread reticence when we learn that it is now possible to add a bionic arm and thus change fundamental "natural" human structure or to perceive the world when the "natural" retina no longer does so. In spite of its great faculty for plasticity, can the brain tolerate unlimited additions to its "natural" structure and still manage to control them? In this field, as in biotechnology in general, we still need to define the scope of clinical, social, and fundamental applications of BMI. By replacing "natural" parts of our bodies by bionic apparatuses bursting with electronics and mechanical gadgets, is man gradually replacing *Homo sapiens* with a new species?[31] For certain prophets of doom, the arrival of cyborgs would precipitate a regression of our species.[32] In 2004, Kevin Warwick at the University of Reading,[33] himself a cyborg in that he has a microprocessor implanted in his forearm, expressed his enthusiasm about advances in cybernetics but indicated the social and psychological fears that microprocessors might provoke if directly attached to the brain. Chris Crittenden of the University of Maine concluded that cyborgs could incite humanity to engage in *self-deselection*: "Our technologically based culture is the first step in the descent toward self-deselection. Although many scholars see positive uses of the cyborg imagery I argue that they downplay, or in many cases entirely ignore, the dangers. Dangers that, if they come to pass, are apocalyptic." So, are we at the dawn of self-deselection, or are we entering an era in which humanity will take full control of its environment? Everyone will have his own answer.

Robots with Neurons

Human beings are not the only ones to benefit from the technological prowess of biomedical research. There also exists a robot functioning with a tiny biological brain, whose neurons are able to learn simple behavior, like avoiding an obstacle. It was developed by Kevin Warwick's team in Reading. Called Gordon, it has a biological brain made of nerve cells from a rat. After collection, these neurons were dissociated then applied to a substrate on some sixty electrodes. After a few hours they established new contacts among themselves and, after twenty-four hours, a complex neuronal network formed in vitro. Seven days after being maintained in this artificial environment the neurons showed spontaneous discharges similar to those seen naturally in an awake brain. The trick was to link the electrodes on which this small cluster of neurons was growing to the motor commands of the robot. From then on, Gordon was under the control of this tiny rat brain, which began to learn certain tasks. For example, when it ran into a wall, the brain received this information as a stimulus and learned by repetition to avoid the obstacle. Certainly, the repertoire of learned behavior remained very limited, but the team estimated that scarcely a hundred thousand neurons were active in this minute brain: Gordon's technical prowess must be compared to that of a rat, whose brain contains a few ten million neurons, or man with several tens of billions. By trying to increase the number of surviving neurons the team hope to one day see Gordon accomplish much more complex tasks than simply avoiding an obstacle.

Fiat Lux

A new biological discipline, *optogenetics*, was born a few years ago. It combines genetics with the tools of modern optics to reveal the relationship of nerve circuits with specific behavioral tasks. This was first realized in 2002 by Gero Miesenböck, who was working at the time at the Sloan-Kettering Cancer Center in New York on *Drosophila*.[34] This approach was extended to mammals by Karl Diesseroth and his colleagues at Stanford University.[35] Both groups contributed to showing how molecular biology could combine with optical methods to activate or inhibit a group of neurons with light, like a switch, and at the same time influence the behavior of an organism, whether a worm, a fly, or a mammal. Today, by controlling light with fiber optics or lasers, this technique enables mammalian brain function to be controlled with unequalled precision in time and space. This capacity constitutes a real revolution, in that light previously was mainly used for observation, from classic microscopic study of the shape of neurons to fine movements of identified molecules in a cell.[36]

To understand the logic of this hybrid method between genetics and optics, we must first recall the principles of neuronal function. Neurons are excitable cells; that is to say, their membrane is selectively permeable to certain ions, such as sodium, calcium, potassium, or chloride. Ion flow leads to changes in charge, which polarize the membrane and produce local potentials or propagated action potentials, which form the nervous impulse. Neurons receive and pass on information in these impulses. This information is transmitted via electrical junctions or by the activation of receptors to neurotransmitters. The principal excitatory transmitter in the brain is glutamate. It is released at the synapse, the zone of connection between neurons. Once released, glutamate activates receptors on the surface of the postsynaptic neuronal membrane, which causes the movement of positive charges carried by cations (positively charged elements such as sodium, calcium, and potassium) to the inside of the neuron. If this depolarization is sufficient, the threshold of excitability of the neuron is reached, triggering an action potential. Because of the electrical properties of the neuronal membrane, electricity has been commonly used to stimulate neurons or synapses. A major inconvenience of this method is the absence of specificity, as current delivered to nervous tissue propagates to nearby excitable structures. To overcome the problem scientists have sought ways to avoid electrical stimulation. The discovery of receptors able to be stimulated rapidly and reversibly by light has offered the possibility of using light as a stimulus instead of electricity. Photoreceptors using rhodopsin in the retina or phytochrome in plants are examples of light-sensitive receptors. But an ion channel expressed by a prokaryote cell soon attracted neuroscientists.[37] A protein permeable to cations is a light-sensitive receptor in the single-celled green alga *Chlamydomanas reinhardtii*. It is called *channelrhodopsin* because of its relationship to retinal rhodopsin. Its maximum light absorption is at a wavelength of some 440 nanometers, like the blue light of the ocean depths where these algae live. This receptor allows the algae to control various cellular functions like intracellular acidity, calcium balance, and electrical excitability. The direct coupling of light to the channel depolarizes (makes more positive) the membrane potential rapidly and strongly, two parameters that have facilitated the use of this protein in neurobiology.

There is no point in exciting neurons if they cannot be inhibited. There is a proton pump (selective to protons and chloride, a negatively charged ion) activated by yellow light of a wavelength around 584 nanometers. This membrane pump, called *halorhodopsin*, was isolated and characterized from the Saharan archaeon *Natronomonas pharaonis*. When introduced into a neuron it causes the membrane potential to decrease when exposed to yellow light, the opposite effect to channelrhodopsin, which causes inhibition. Thus channelrhodopsin and halorhodopsin are switches with which the experimenter can respectively excite or inhibit a given neuron using a light beam in the blue or yellow ranges of wavelength.[38] Because these two channel systems can be transferred genetically to neurons or other excitable cells, this ability to use light to activate or inhibit neural circuits could have practical applications in clinical medicine, such as for Parkinson disease.

The intrinsic properties of proteins of the rhodopsin family (rapid and sustained activation by light and fast closure of the channel in the absence of light) make them choice molecules to modulate at will the membrane potential and thus the state of activity of neurons. Because the green algal protein is not naturally present in neuronal membranes, scientists use genetics ("optogenetics") to force its expression in target cells. By using viral vectors containing complementary DNA coding for rhodopsin one can cause this light receptor to be expressed in a specific population of neurons.[39] Once infected by the virus the neurons respond to a flash lasting a few milliseconds by a depolarization that, if sufficient, can cause the neuron to discharge.[40] This light stimulation is reversible, and at the end of the flash the neuronal membrane potential returns to normal, and the impulses cease. Today light stimulation is beginning to supplant the electrical stimulation that enabled progress in neurobiology from Volta, who in 1792 described the excitation of frog muscle, to Penfield and his famous homunculus (see chapter 3) traced by electrical stimulation of the cortical surface. One knew, for example, that one microamp of electrical stimulation of the primary somatosensory cortex could elicit sensations comparable to those caused by tactile stimulation of the periphery,[41] but such cortical stimulation did not permit a measure of quantity or quality of neurons at the origin of the sensation.

In 2008 optogenetics was used for microstimulation by light to activate a group of cortical neurons.[42] The declared object of this experiment was nothing less than to define the cellular basis of perception. Stimulation of only three or four neurons in the somatosensory cortex of a mouse was enough to trigger a sensation similar to stimulating a single whisker. The stimulation was not at the periphery but by a miniature light-emitting diode (LED) on the cortical surface. The mouse received a flash that directly stimulated cortical neurons expressing channelrhodopsin. The number of neurons recruited by the stimulus was proportional to the light intensity. How can one know that the mouse really perceived the flash as a true sensory stimulus? By submitting it to Pavlovian conditioning.[43] Mice were trained to associate light stimulation of the cortex with a reward, a drop of water, delivered only when their snout was in the correct orifice of an apparatus. At the start of the test the mouse placed its snout in a central cavity, then received five brief flashes every fifty milliseconds to the primary somatosensory cortex. After this series of stimuli the mouse put its snout in an orifice on the left, or on the right if there had been no stimulus. After a few training sessions all the mice expressing rhodopsin achieved a correct response figure of eight out of ten. To reach this level, at least sixty neurons needed stimulating. If only one flash was given instead of five, inducing a single electrical impulse in the stimulated neurons, almost three hundred neurons needed recruiting simultaneously to obtain the same success rate. These experiments showed that the formation of a percept necessitated the activation of sixty neurons each firing simultaneously five impulses, or three hundred neurons each firing a single impulse. Surprisingly, these experiments suggest that neural circuits can choose to function collectively or individually. In any case a relatively small number of active neurons is necessary to construct a percept and allow an animal to make a correct decision.[44]

At the Pasteur Institute we have recently shown that new neurons born in the adult brain can be stimulated by light.[45] We used optogenetic techniques to make the new neurons sensitive to light. The specific expression of channelrhodopsin in the new neurons enabled us to control their electrical activity simply with light flashes. For the first time this method revealed the signals emitted by the new cells and identified their targets in the olfactory bulb. With progress in the use of light to excite or inhibit neural circuits,[46] it is certain that optogenetics still reserves many surprises for us in understanding how cortical circuits relate to different aspects of behavior. *Fiat lux*!

Botond Roska of the Friedrich Miescher Institute for Biomedical Research in Basel also took advantage of optogenetics by introducing halorhodopsin into the retina of blind mice. This work, in association with José-Alain Sahel's team at the Vision Institute in Paris, used mice with an incurable genetic degenerative retinal disease corresponding to human retinitis pigmentosa or, to a lesser degree, age-related macular degeneration. They aimed to restore visual function by introducing a light receptor in surviving neurons. Because the disease involved degeneration of the photoreceptors, they had to introduce the light receptor in surviving retinal cells which did not express the light receptor. Using a virus, halorhodopsin was introduced in residual cone photoreceptors that had lost their photosensitivity. The blind mice recovered part of their light sensitivity.[47] The same group also showed that human photoreceptors, removed post mortem from retinal explants, could be treated similarly to increase their light sensitivity, thus opening the door to human clinical trials.

The applications of optogenetics go far beyond just repair of the retina. For example, Karl Deisseroth and his colleagues targeted different types of cerebral neuron: such precise targeting might help us understand the neurological basis of psychological disorders.[48] By aiming at the hypothalamus it is possible to simulate vital needs and desires in animals.[49] Optogenetics is also at the heart of research into addiction. Light stimulation can control neural circuits involved in dependence to drugs. Freely moving mice received light flashes to the brain when they entered an area called the "reward room." In a few sessions the mice learned to spend most of their time in this room, carefully avoiding other chambers in the experimental apparatus.

Today optogenetics is becoming a technology for treating human psychological disorders. This technological upheaval raises a number of questions. What will happen when genetic manipulation and miniature electronic devices for delivering light will allow us to stimulate certain regions of our brain at will? What will we do with the knowledge that we can target the hypothalamus and make a subject hungry, angry, or excited? Of course, there already exist patients with implanted electrodes to connect them to a computer, for example, to treat recurrent epilepsy. We described in chapter 5 a relatively new procedure to stimulate deep brain structures, such as the subthalamic nucleus, to treat Parkinson disease. But now this disease may begin to benefit from the miracle of light. In recent research in a Parkinson disease model Karl Deisseroth and his colleagues used optogenetics to study, in real time, circuits that are active in the awake rat when it is planning voluntary movements.[50] Then they reproduced the same effects by stimulating with light different types of neuron in different motor regions. The light stimulated directly neurons of the motor cortex connected to the subthalamic nucleus, thanks to which neuronal activity in the basal ganglia could be reestablished.

Brain Doping

We easily accept the idea of wearing spectacles, smoothing an ugly wrinkle with Botox, reshaping our nose, or buying help with our children's schooling to improve their intellectual ability. Whether physical or mental, we consider that an inherited or acquired genetic imperfection can be treated like any other event, vital or not. If these choices raise no ethical problems, what about medication that we divert from its therapeutic purpose to increase the productivity of our brain? Drugs for stimulating intelligence or memory already exist. We can use medication outside its therapeutic context to improve our mental function or increase our intelligence. Even so, are we ready to accept that people use cognitive stimulants and psychoactive drugs? Is brain doping acceptable in society any more than in sport?

These questions were at the heart of a debate between neuroscientists and philosophers chaired by Judy Illes and sponsored by the National Science Foundation and the New York Academy of Sciences. This meeting considered the problems of the nontherapeutic use of new technology. Today fundamental new progress in neuroscience and its technological applications offers everyone an arsenal of tools for improving motor function, perception, attention, and memory. Possibilities of improving brain power are multiple, whether based on medication, implants, or neuroprostheses.[51] Aimed at improving individual performance, recent progress raises new questions about the respect of moral values and legal rules, and even about individual health. Should we anticipate dealing with the consequences when the brain is involved? The aim of the meeting was also to put into perspective recent progress in technoscience and anticipate social, moral, and judicial implications.[52] It is indeed time to reflect on the limits to impose on the use of new technology, notably the cognitive enhancers called *smart drugs*.[53] By acting directly on the brain and seeking to "enhance" it, that is to say, make it work better than normal,[54] we are perhaps meddling with man as a human being. The debate is open. While philosophers bring replies, scientists try to overcome the problem. Can the two points of view be reconciled?

The debate about the illicit consumption of drugs prescribed for specific pathology, such as attention deficit or narcolepsy, is open, and opinions are divided. For some the possibility of improving intellectual performance is a potential source of discrimination and thus inequality. There would certainly be those who could afford treatment to improve their intellectual faculties while others would remain without the means of access to drugs to enhance their ability. Are we heading for a two-tier society where a wealthy handful could afford to dope their brain while the rest remain cut off from new knowledge and intellectual prowess? If so, this doping would represent an important and dangerous discriminatory factor. What would happen if some scholars or students doped their brains to improve their performance at school while the rest of the class did not? Would their teachers encourage the whole class to participate? Even worse, for the detractors, cerebral implants, neuroprostheses, or implanted memory chips would lead to a modification of the whole or part of human identity. For example, to satisfy economic needs it would be possible to impose socialized management of our mental activity. We would lose free will and submit to the dictatorship of an instrument. According to this point of view, quintupling man's mental function would mark the start, under the growing influence of cyberneticians, of a radical transformation of man as defined by the humanists of the Enlightenment.[55]

On the contrary, others are convinced by the advantages of technology in the service of mankind and encourage the development of cerebral doping. For these partisans *Homo sapiens* has always succeeded in increasing his mental faculties by discovering the virtues of coffee at breakfast or a cigarette to stay awake late into the night. Molecules produced by the pharmaceutical industry to dope intelligence or memory would only form part of the many ways to achieve this end. For better or for worse, Ritalin, modafinil, and amphetamines have a bright future. Consumption of these smart drugs and psychostimulants in general is on the increase. For example, to treat psychiatric disorders in children and adolescents Ritalin is prescribed without hesitation, notably for treatment of ADHD, as we saw earlier. These children experience significant improvement. Modafinil can increase attention and memory and also reduce time spent sleeping. Its structure and mode of action are different from psychostimulants. It aims to prolong waking and is widely used to treat hypersomnia (excessive sleep), narcolepsy (sudden bouts of falling asleep), and by night-shift workers. It targets the locus coeruleus in the brainstem.[56] Today it is used deviously to overcome sleep deprivation in certain professions, such as the military, firefighting, and nursing. There are numerous websites vaunting its merits and stating that its efficacy approaches that of amphetamines, with side effects "more like caffeine."[57] At the time of its marketing in 2002, the press observed that Provigil (a brand name for modafinil) "is showing signs of becoming a lifestyle drug for a sleep-deprived 24/7 society."[58] BBC News announced that a study on sixty healthy volunteers showed that it significantly improved short-term memory.[59] Maybe this unexpected popularity means that coffee and cigarettes have reached their limit and can no longer satisfy workaholics. If so, perhaps doping the brain is a subterfuge to enable modern slaves to stay in the race toward ever more competition, like the Red Queen in Lewis Carroll's *Through the Looking-Glass*.[60]

Alongside the psychostimulants is another category of new substances to increase attention and vigilance and help learning and memory. Called *nootropics*, they were originally conceived for patients with memory problems. Today they figure among the chemical weapons used by some students trying to improve their intellectual performance. Because of their ability to improve cognitive capacities these new drugs are set to develop wildly over the next few decades. Nootropics will mainly be consumed to fight attention lapses among tired workers or by travelers needing reinvigorating after a long flight. Examples of these molecules are the *ampakines* developed by Gary Lynch and his colleagues in California. They facilitate conduction of the nerve impulse by activating glutamate receptors.[61] They were developed for people with chronic fatigue or Alzheimer disease, but the main consumers now are healthy people wanting to boost their memory. Unlike psychostimulants like caffeine or amphetamines, ampakines seem not to provoke lasting unpleasant side effects such as insomnia.

Narcotics are addictive substances like morphine, cocaine, or cannabis. Cocaine and amphetamines[62] increase wakefulness by stimulating liberation of dopamine, the pleasure molecule. Psychotropic drugs literally "act on the mind." In 1957 Jean Delay, the director of the Institute of Psychology in Paris, defined them as natural or artificial chemical substances that act on the mind and are susceptible to modify mental activity, although he did not specify the nature of the modification. All psychotropic drugs cause dependence (in the DSM-5, referred to in chapter 5, replaced by the term addiction),[63] whether they consist of amphetamines, neuroleptics,[64] or hallucinogens, like mescaline extracted from a Mexican cactus formerly used by tribes in Mexico and the southwestern United States in magico-religious rites. This category of substances acts on neural reward circuits to produce a dopamine effect.

Amphetamines are synthetic substances that stimulate the central nervous system. They were introduced in the 1930s as nasal decongestants, when they were available over-the-counter as Benzedrine inhalants. Later they were used to treat obesity, then mood disorders. In the 1950s they came into widespread use for women wanting to lose weight. Such abuse took on epidemic proportions in the United States as the number of dependent individuals grew. From being prescribed as stimulants, they are now classified as narcotics. They increase the alertness of the major functions of the brain, hence their use among professionals ever keener to better their performance and achieve the highest standards. It is common to find amphetamines in use among financial traders. The down side is the physical and psychological risks: chronic sleeplessness leading to extreme fatigue, hypertension with possible cardiovascular accidents, anorexia, acute delirium, and paranoia. The TREND report of the French Observatory for Drugs and Dependence (OFDT) noted that the consumption of amphetamines existed in France since at least the early 1940s. From limited beginnings, it expanded in the 1960s but without reaching the proportions of some other European countries such as Sweden or the United Kingdom.[65] Until 1995, the supply of amphetamines was essentially from the illicit use of prescribed drugs. Imported amphetamine powder has emerged in response to growing demand from the "techno" scene.[66]

Ecstasy is another substance that is in the news, although the number of consumers is fairly low in France, some 2 percent of people from fifteen to sixty-five years old. Ecstasy causes a large, fast increase in serotonin, a transmitter that regulates mood, pain, and aggressiveness, and in dopamine, the main control over the wanting system discussed in chapter 3. Ecstasy, or probably one or more of its metabolites, blocks recapture of serotonin and dopamine. Sadly, these metabolites might also destroy neurons. Although Ecstasy gives the most pleasure compared with its toxic effects, it still remains one of the most mortally dangerous, and it can cause irreversible psychic problems: a serious aspect of regular consumption of Ecstasy is its probable link to depression and cognitive disorders. Some scientists even think that neuronal destruction by Ecstasy could provoke degenerative diseases such as those of Parkinson or Alzheimer.[67] In addition, Ecstasy is often taken with other substances, and the synergistic or additive effects remain unknown.

This range of drugs demonstrates the chemical diversity and variety of uses of substances taken less for reaching an artificial paradise than for attaining ever more ambitious socioprofessional productivity. In 2008 a group of scientists launched a controversy by pleading for a responsible use of medication to improve mental function in healthy people.[68] If the industry produces substances that can improve mental function without negative side effects in the short or long term, why refuse them? "In a world in which human workspans and lifespans are increasing, cognitive enhancement tools—including the pharmacological—will be increasingly useful for improved quality of life and extended work productivity, as well as to stave off normal and pathological age-related cognitive declines," argued Henry Greely of Stanford Law School and his neuroscientist colleagues. This put the cat among the narcotic pigeons and provoked contrasting reactions throughout the world. According to this group, drugs that stimulate intellectual functions should be recognized and their use authorized within so far undefined limits. After all, is not psychic stimulation the main objective sought daily by a good number of us in our espresso and accepted for ages?

To be more convincing, the advocates of brain doping attack directly the three arguments used to condemn it. The first concerns the bias that this usage could engender at work or in education. In view of the already accessible arsenal of psychostimulants such as tobacco, vitamin C, or coffee, to recognize the use of smart drugs would be a scientific way to validate and control them. The group then attacks the myth of the prowess of "natural" rather than "artificial" learning, by analogy to the world of sport. The difficulty with this debate that has raged since antiquity is to define the borderline between natural and artificial.[69] The authors remind us that writing or informatics that help learning are not natural. Finally, because drug abuse is a social scourge society seeks to control the use of substances because of their toxicity and risk of dependence, but we now know that dependence on heroin, hypnotics, and nicotine shares the same neurological mechanisms.[70] However, tobacco does not have the same legal status as heroin. In practice, the fact that a substance is addictive is not enough to ban its use.

The consumption of unprescribed cognitive stimulants is already a reality, as Greely and his colleagues remind us in the introduction to their arguments. According to various enquiries, on American university campuses more than a quarter of students obtain drugs such as modafinil to stimulate vigilance or Ritalin to increase performance before critical examinations. As we age, we are confronted with normal diminution of memory or attention; should we not use a stimulant to recover our former abilities? To ask this question *Nature* accompanied the article by a survey of 1,400 scientists. One in five admitted having already taken Ritalin, modafinil, or beta blockers (designed for treatment of cardiac dysfunction) to improve attention or memory or simply to control stress. The survey showed that this phenomenon is far from limited to students preparing for their examinations: maximum consumption affects the over-fifty-fives as well as the under-twenty-fives. This tendency is also reported in other professions, such as show business or politics.

The question of brain doping would not be complete unless we addressed the use of transcranial magnetic stimulation (TMS) to improve mental faculties. Indeed, we have recently learned that training the memory is more effective if associated with TMS. Research in Tel Aviv demonstrated that with a helmet delivering TMS to specific brain areas, cognitive function in patients with Alzheimer disease could be improved.[71] So the same questions as for psychostimulants are relevant for this technique.

The Infinitely Small in Our Brain

Some believe that Lilliputian engineering might be useful to manipulate or repair the complex machinery of our brain. At the limits of the infinitely small are hidden physical and chemical laws necessary for the brain's unique properties. Could there exist a common program linking the technosciences of matter, life, and informatics? Clearly, the answer is yes, and we are already inundated by commercial nanomaterials in such diverse forms as carbon nanotubes, nanolasers in our DVD players, nanochips for identification and traceability, and even tools for diagnostics. Under the heading of "nanotechnology" are grouped scientific and technological activities at an atomic or molecular scale and the principles and new properties that emerge at this scale.[72] The conceptual framework covers a variety of disciplines such as physics, chemistry, and biology, with the subject of study being minute objects with singular properties. At their origin, the nanosciences shared the common objective of mastering the synthesis and studying the behavior of nanoelements in different environments. Among other things, the fundamental principles of nanoscience predict that tomorrow's drugs will be more efficient once transformed into nanoparticles. When this still nascent technology is fully mastered it will probably cause upheavals that we can scarcely imagine today, as much in economic and social spheres as in technology. This revolution, already under way, will certainly surpass the upheavals caused by the industrial, mechanical, and computing revolutions of the previous two centuries, and in a record time! Although it is still impossible to predict when in the more or less near future the nanotechnological revolution will change our environment and quality of life durably, it seems inevitable.

What exactly will this science, born of the study of the almost infinitely small, bring us? Nanoscience deals with physical particles of the order of a nanometer in size, a billionth of a meter. This dimension was for long a field neglected in favor of more classic micrometric scales, a millionth of a meter. Nanoscience involves sizes just greater than that of the atoms and molecules that form the building blocks of inert and living matter. The mean size of an atom is about a third of a nanometer: in one nanometer we could theoretically line up three atoms. A sphere of two or three nanometers in diameter contains just a few atoms. We can therefore conclude that nanometric objects contain just a handful of atoms or molecules, unlike classic macroscopic objects, which contain an astronomic number.

The physicochemical properties of nanoparticles are part of the application of fundamental knowledge to industrial and medical applications. For example, their properties allow the transfer of active principles directly to the inside of a diseased cell. However, to be useful nanometric elements can only be handled once inside larger, micrometric, vectors, themselves inside something that can be handled by man. For this reason nanoscience and nanotechnology also involve the manipulation of larger, microscopic, objects: they do not simply exploit the properties of atoms or molecules to obtain a result at the macroscopic level. Even if the work is done on the nanoscale, the final result is at a higher level. In this tiny world quantum mechanics rules supreme. Control and manipulation of quantum states open the way to methods of calculation far beyond what our current computers can handle. This field could not have developed so spectacularly these last few years without improvement in observation methods and the development of new instruments like the *scanning tunneling microscope*, designed by Gerd Binnig and his colleagues in 1985 in IBM's labs in Zurich.[73] This microscope is one of the most fascinating recent tools in that it allows matter to be probed at the level of the atom, thus revealing a previously inaccessible world.[74]

In 1959, the Nobel laureate Richard Feynman gave a talk to the American Physical Society at the California Institute of Technology entitled "There's Plenty of Room at the Bottom," which is considered the inaugural lecture in the history of nanotechnology. He evoked a research field that had remained undeveloped, that of the infinitely small. He considered the possibility of concentrating large amounts of information on minute surfaces by manipulating atoms, just as others manipulated cinderblocks. For example, he calculated that by reducing each printed dot to a circle one thousand atoms in area one could write the whole of the *Encyclopaedia Britannica* on the head of a pin. This claim is still widely cited for, even if at the time it was possible in theory but not in practice, progress in precision technology make this challenge possible. But the real epistemological rupture proposed by Feynman is in the manipulation of individual atoms and molecules as discrete elements rather than in continuity.[75] Fifteen years later, in 1974, the term nanotechnology was suggested by Norio Taniguchi of the University of Tokyo to describe research with the objective of synthesizing and studying very small objects with new properties. To deal with such singularities this technology marks the convergence of three so far relatively independent fields: physics, chemistry, and biology. Modern nanotechnology includes the study of material phenomena at the atomic, molecular, and macromolecular scale, where properties differ significantly from those at a larger scale. Whereas at the micrometer scale the laws of classic physics and chemistry apply, it is not true at the nanoscale, where traditional physicochemical properties change. Nanoscience explores new laws applicable to nanometric elements.[76] Nano-objects are intermediate between basic building bricks of matter (atoms and molecules), which are obviously discontinuous, and continuous macroscopic matter.[77] This rather special situation gives them dynamics and properties different from the small molecular or macroscopic systems that prevail at larger scales.[78] To be applicable to man nano-objects will have to undergo microscopic or macroscopic transformation, and it will be necessary to develop materials and devices with new functions and performances, that is to say, construct matter at the nanoscale.

Today nanotechnology directly applicable to neuroscience concerns three sectors: nanoelectronics, nanobiotechnology, and nanomaterials. Nanoelectronics derives from microelectronics and concerns components for computers. Ten years ago engineers at the Intel research center at Hillsboro, Oregon, produced transistors twenty nanometers (seventy atoms) in size.[79] Today transistors can be as small as seven atoms long, and according to Michelle Simmons, the inventor of the smallest transistor so far, computers using them could resolve problems that would take longer than the existence of the life of the universe for a classic computer. She claims that the silicon transistor is the first that leads concretely to the manufacture of a quantum computer. She predicts that such a transistor will be used commercially within five years. Developed at the Centre for Quantum Computation and Communication Technology of the University of New South Wales and the University of Wisconsin at Madison, this transistor could reduce the size of the processor a hundredfold and accelerate its speed beyond our wildest dreams. When this technology arrives, the famous Moore's Law will be brushed aside.[80] The possibility of producing microprocessors with prodigious calculating speed allows us to envisage online processing of brain waves and interfering with thought.

Nanobiotechnology uses living cells or isolated molecules such as genes or enzymes as tools to manipulate living organisms or construct materials inspired by biological systems (one speaks of biomimetics). Possible fields of application in matters of public health include, for example, biological implants, new diagnostic tools, and specific targeting of diseased cells, as we discussed earlier.

Nanomaterials depend on precise control of the shape of substances or particles at the nanometric level to build "nanostructured" materials. An oblong-shaped active principle may be transported efficiently by the cardiovascular system; a spherical form may favor its fusion with cell membranes and thereby an intracellular effect. Nanomaterials are composite structures that exist in various forms, free or fixed, fibers or tubes, crystals or lamellae, with specific properties related to their large surface-to-volume ratio: reliability, adaptation, and resilience. For example, they easily penetrate the body not only via respiration and digestion but also through the skin. They can circulate in the blood and thus easily reach the brain. They can also reach the brain through the nose and olfactory nerves.

Foreseeable medical and industrial applications are so numerous that there is little risk in predicting their penetration of health and economy in the near future. In 2015, 15 percent of the world's manufacturing activity should involve devices or materials benefiting from nanotechnological progress. According to the National Science Foundation, the world market, estimated at 500 billion dollars in 2008, could double by 2015. However, like all major scientific and technological progress, nanotechnology is not devoid of problems. In spite of the obvious benefits of this technological revolution, numerous questions remain, raising fears in the realms of health, ethics, and even industrial risk. Manipulating matter at the molecular or atomic level and interfering with life involves high stakes for which we must prepare ourselves. For the authorities the greatest risk of all is that public opinion rejects this technoscientific progress en masse.[81]

The Birth of the First Nanoelements

The first nanometric object discovered was *fullerene*. This organic matter belongs to a large family of compounds, composed of carbon, able to form spheres, ellipses, tubes, or rings. Fullerenes are similar to graphite and are formed of a lattice of rings. Fullerene was first produced in 1985 by Harry Kroto, at the University of Sussex, and Robert Curl and Richard Smalley, at Rice University. They received the Nobel Prize in Chemistry in 1996. This compound is composed of sixty carbon atoms and was named in honor of Buckminster Fuller, the American architect who designed a geodesic dome for the Montreal Expo '67 made of hexagons and pentagons arranged into a sphere. These three scientists certainly did not expect to discover a new molecular form of carbon. Before their discovery, organic chemistry books noted the existence of two forms of carbon: graphite and diamond. The first fullerene discovered consisted of twelve pentagons and twenty hexagons, each vertex corresponding to a carbon atom and each edge a covalent bond. This truncated icosahedron is similar to the structure of a football, hence the term footballene.

In 1991 Sumio Iijima discovered that these elements could exist as carbon nanotubes. These tiny tubes have remarkable mechanical and electrical properties, which is especially useful in neurology, where materials with good electrical conductance are needed.[82] Fullerene science stimulated the development of carbon nanotechnology and its important role in future electronics and biomaterials. The emergence of these tools is indicative of the about-turn by chemistry in the twenty-first century. Today nanochemistry is legitimized to synthesize structures of unimagined complexity and architecture. Chemistry seems to be leaving the domain of exploration to become a creative science. New methods of biomimetic synthesis will allow the creation of unknown structures based on various atoms including metals. They may enable the creation of structures with properties of autoassembly and auto-organization leading to complex macromolecules. The possibility of equipping a soldier with an exoskeleton to protect him on the front lines might be one application of new nanochemical technology.

Nanotechnology and Medicine

Medical progress over the centuries has allowed diagnosis to become precise and early. In the beginning, the doctor had to be satisfied with a clinical examination of his patient. Later, imaging allowed the examination of organs or parts of organs. Biopsy allowed the analysis of tissue samples and the establishment of data banks. Today, observation and manipulation at the subcellular level can identify the origin of pathology and suggest procedures to cure it. Tomorrow, molecular imaging will study the behavior of single molecules within "organized molecular systems."[83] Use of nanoparticles as contrast agents for MRI is a concrete modern example of a benefit of nanotechnology in medicine. Nanomedicine will help establish a molecular diagnosis and then use therapies to target single molecules. Before a disease becomes detectable there are physical and chemical changes in molecules: detecting such precursors would allow prevention rather than cure, the attempt at which often comes too late. In 2010 a group at Harvard University, trying to satisfy this criterion of anticipation, demonstrated the possibility of introducing a nanometric transistor into a cell to measure its electrical activity.[84] This work led to implantable biochips to check one's state of health online and even deliver any necessary medication.

For tetraplegics biochips can provide brain-machine interfaces. In one approach aiming to mimic biological properties of sensory systems nanotechnology has allowed the creation of an electronic nose. Nano-receptors translate the chemical signal from organic molecules in a gaseous environment into electronic signals analyzed by microprocessors much as in the neural circuits of the olfactory system. The detection system may be a semiconducting nanostructured surface of which the conductance is modified by interaction with an organic molecule. Such bio-inspired systems for the detection of odoriferous substances have many implications for narcotics and mine detection and in the food and chemical industries.[85]

The ever more frequent use of *quantum dots* to mark and trace cell molecules by replacing fluorophores also indicates a radical change in classic pharmacological concepts.[86] These semiconducting particles of known diameter emit light at a given wavelength, unlike classic fluorophores. By varying the size of the nanoparticles one can obtain a palette of colors. Once fixed on biologically important molecules these nanoparticles can be followed in living cells or organisms and may orient future therapy toward specific actions at the molecular level.[87]

Today, one of the most promising fields in nanotechnology remains the possibility to vector future drugs to increase their specificity. A drug is only worth its ability to reach its target, to be in the right place at the right time at the right concentration. So the ideal drug should not be lost in the complexities of the human body, to be diluted and lose part of its efficacy. To this end the active principle might be wrapped in a tiny cage a few nanometers in diameter and made of structured polymers. Protected by this shell tomorrow's drugs will journey unharmed through the organism and reach their targets. Because the diameter of blood capillaries is about 4 micrometers and their pores about 50 nanometers, modern research seeks to produce nanovectors able to cross the pores to hit their target. Some trials involve encapsulation inside nanometric structures such as spheroidal fullerenes or carbon nanotubes; others involve magnetic nanoparticles guided from outside the organism using a magnetic field aimed at the area to be treated. To be effective the vector must liberate the active principle once it reaches the target.

This could be automatic, after contact with a specific molecule or detection of a particular physicochemical environment such as a change in pH or temperature, or remotely controlled by ultrasound, magnetic field, or infrared to raise the temperature. Nanocapsules that deliver drugs on demand would join the nanotechnological therapeutic arsenal. As an example, German scientists produced a polymer chip with microcavities containing a soluble substance.[88] Each microcavity was sealed by a cover that could be destroyed by an electrical current, thus liberating the active principle on demand. Recently the same team modified their approach using nanotransistors to stimulate neurons.[89]

Could gene therapy, consisting of introducing genetic material into cells to correct a pathological anomaly, benefit from nanotechnology? Two young infants suffering from a rare X-linked severe combined immunodeficiency disease (X-SCID) recently benefitted from a medical "first" by the team of Marina Cavazzana-Calvo and Alain Fischer at the Necker Hospital for Sick Children in Paris. A harmless retrovirus, which penetrated cells spontaneously, was used as a vector for the gene, which was integrated into their genome. Today nanoparticles are becoming the new vectors, replacing viruses by imitating the viral capsid (the envelope of the virus containing the genetic material). Recombinant nanoparticles are not infective because they have no viral genetic material and thus do not replicate in target cells: their biological effects cease at the end of the treatment.

The effectiveness of nanomedication is also demonstrated by therapy of resistant liver cancer or leukemia. In these cases a natural lipid, squalene, is coupled to anticancer drugs to form nanoparticles with decoupled efficiency. Success is also expected in the development of nanomedication able to cross the blood-brain barrier, which partially isolates the brain from the blood stream.[90] This concerns many diseases including Alzheimer and Parkinson, as well as stroke. Thanks to miniaturization, onboard diagnostics, and improved coupling between electrodes and tissues, nanotechnology is developing better performing implants. Progress in nanobiotechnology is opening the way for new drugs, implants, and nanosystems and almost infinite improvements in human physical and mental performance.

When Nanoscience Merges with Life

The meeting of hard science, like physics and chemistry, with biology is not new. The beginning of the twentieth century already saw the fruitful merging of chemistry and biology to give birth to biochemistry: chemistry in relation to life. This alliance produced remedies not only from natural extracts of plants but also from chemical synthesis, thus leading to an almost infinite increase in the number of possible drugs. The second half of the twentieth century saw physical methods applied to living organisms. This convergence led to molecular biology, of which one of the first notable successes was the resolution of the three-dimensional structure of DNA by X-ray diffraction. Today, we are witnessing the transfer of theoretical nanoscientific knowledge to practical applications. First microtechnology and now nanotechnology have revolutionized informatics and communication. Our everyday experience is that the power of our personal computers doubles about every two years. If this progress is maintained, in a few years we shall all possess a portable computer of which the calculating and storage capacity will be more than that of the whole planet in 1950! Such progress is possible thanks to the extreme miniaturization of electronic components and the continual size reduction of transistors. In 1950 a transistor had a size of a few centimeters (10^{-2} meters): today a transistor is a few nanometers (10^{-8}) in size, a million times less. Modern technology implies manipulating matter at a resolution between a micrometer and a nanometer. Given that the two operate at similar scales and sometimes share common laws, it is natural that nanotechnology encounters biology to form *convergent* technologies. We shall not attempt an exhaustive definition of what these are or are not.[91] Indeed, from various reports it emerges that there is no consensual definition. Some people see therein the mark of an emerging, developing discipline. For example the National Nanotechnology Initiative (NNI) defines nanotechnology as science, engineering, and technology conducted at the nanoscale. This definition is too restrictive because it does not consider devices to manipulate objects or fluids at the microscale nor macroscopic structures that contain nanostructures.

The fusion of nanotechnology and biology can take two different routes, using contrasting strategies. The *top-down* approach reduces objects measured in centimeters or millimeters to the nanoscale while conserving their function. On the contrary, the *bottom-up* method relies on growth in size from nano-objects (atoms or groups of atoms) to form new macroscopic objects. Today, the top-down approach dominates most nanobiotechnology. The bottom-up approach remains limited, for it depends on the development of new theoretical concepts to predict emergent properties when passing to the higher scale, and this only from knowledge of individual properties of component units. Faced with growing therapeutic needs in Western society precipitated by various factors, notably demographic evolution and an ever-increasing life expectancy, the contribution of nanotechnology to public health is vital. The ever more pressing demand for new remedies is encountering major obstacles because our resources are not indefinitely extensible. We must add to this exhaustion of resources a penury of ideas that is currently hitting the pharmaceutical industry, leading to stagnation in therapeutic innovation. So there is motivation for new in vitro tests using miniature devices to analyze more quickly the effects of candidate drugs at the molecular level.

One of the most rapidly developing applications is tissue engineering. It involves three important stages:

(1) grafting autologous cells (or allogeneic or xenogeneic cells after in vitro amplification);
(2) implantation of tissues reconstructed in vitro from cells and molecular scaffolds (such as polymer hydrogels);
(3) regeneration of tissues in situ using intelligent biomaterials.

Millions of people throughout the world, including fifty million in Europe, have already benefitted from tissue engineering, which has been used to replace, repair, or improve a defective organ. These artificial tissues go from cell aggregates to complex tissues able to replace an entire organ. Thanks to nanotechnology and microfabrication borrowed from the semiconductor industry it is now possible to place cells with unequalled precision and control their behavior. By combining research efforts in nanotechnology, stem-cell biology, and tissue engineering, skin and cartilage substitutes have been grafted on thousands of people.

Today, tissue engineering seeks to reproduce the internal structure of a given tissue as faithfully as possible, for cell function depends closely on the microenvironment. With the help of new techniques and materials specialists aim to create hybrid materials combining nanostructured materials (organic polymers or minerals) and living cells to replace defective tissues. The major challenge the engineer faces is to produce biocompatible materials that blend with neighboring tissues without being rejected. Reconstruction of an artificial cornea from recombinant human proteins from the extracellular matrix and epithelial, stromal, and endothelial cells from adult stem cells is an example of recent success in tissue engineering.[92] Numerous other examples of reconstruction of epidermis, blood vessels, or nervous tissue illustrate to what extent nanotechnological methods today find their expression in life science. Not all organs can yet be directly affected by tissue engineering, but it is certain that in the near future these new tools will permit better diagnosis and treatment of defective vital functions. Is the Grail within reach?

Neurofeedback: Cerebral Bodybuilding

The objectives of neurofeedback in clinical practice are simple. The aim is to give the patient control over his mental activity, including certain unconscious functions, in order to prevent or treat psychiatric or neurological disorders.[93] The development of neurofeedback began in 1875 with the discovery of electrical phenomena on the surface of animals' skulls by Richard Caton (1842–1926). Using a galvanometer to detect electrical activity, he measured fluctuations of evoked potentials in different brain regions during sensory stimulation. After Caton's death, Hans Berger continued similar experiments on human subjects. Berger recorded a human EEG for the first time in 1924. He recognized oscillations at different frequencies and described *alpha* waves (from 8 to 12 hertz).[94] With great intuition, he postulated a close relationship between mental function and variations in the EEG rhythm. He proposed that anomalies at certain frequencies could allow psychological or neurological disorders to be characterized.[95]

In 1968 Joe Kamiya at the University of Chicago popularized neuro-feedback through his work on alpha rhythm. In one of his famous studies subjects were asked to close their eyes and concentrate on sounds. Gradually they learned to recognize their own alpha waves during the sound stimulation. A second stage showed that once they had learned, subjects could reproduce alpha waves without the sound. These observations popularized neurofeedback in the United States in the 1970s. We not only have recourse to drugs to manipulate our brain's chemistry, but we can also use various technological processes to modify its physical properties. By simply listening to specially designed sound sequences or using devices emitting light signals at a specific frequency, we can choose the waves that dominate our brain rhythms and even synchronize the electrical activity of our two hemispheres.

A few years after Kamiya's experiments, Barry Sterman and his colleagues in Los Angeles demonstrated that the effect of learning on the genesis of our brain rhythms went beyond the simple regulation of alpha rhythm. They showed that we could also control the production of beta waves.[96] They noted that alpha wave frequency was 8 to 12 hertz and their amplitude 25 to 100 microvolts (amplitude represents the intensity of the waves). Alpha waves were mainly in occipital regions and were weaker more anteriorly. On the other hand, beta rhythm had a frequency from 13 to 30 hertz and a smaller amplitude (5 to 15 microvolts) and were principally in sensorimotor cortical regions. Opening the eyes eliminated alpha rhythm instantly, though beta waves persisted. Sterman's team showed that part of the beta regime (slow beta waves, at 13 to 16 hertz) were present in the sensorimotor cortex of the awake animal and during certain phases of sleep. Unexpectedly, they discovered that monkeys and cats trained to control the appearance of sensorimotor cortical rhythms by Pavlovian learning were relatively resistant to epilepsy. So, fortuitously, the first application of neurofeedback was established. These results incited Sterman to use neurofeedback to treat patients suffering from frequent epileptic attacks. The adventure was just beginning.

In view of Sterman's growing success, soon reproduced by other teams, a new scientific society was formed: the International Society for Neurofeedback and Research. Their objective was to share the ever more voluminous collection of data from numerous EEG laboratories. A database with results from many parts of the world was set up to compare EEG results from healthy and diseased subjects. This globalization of scientific data helped neurofeedback extend to varied neurological and psychological pathology. Over the next decades it was used to treat not only epilepsy but also migraine, chronic pain, insomnia, depression, obsessive-compulsive disorder, drug and alcohol addiction, and post–traumatic stress disorder (PTSD).

PTSD is extremely disabling and is characterized by three features: reliving the traumatizing event in the form of flashbacks in the day and traumatic scenes in nightmares; avoidance by the victim, employing considerable efforts to eliminate thoughts or situations associated with the trauma; and hypervigilant anxiety, with the victim remaining on permanent alert.[97] In 1989 Eugene Peniston and Paul Kulkosky at the Veterans Medical Center at Fort Lyon successfully used neurofeedback to treat veterans suffering from severe alcoholism. Two years later they widened their field of application to PTSD in veterans of the Vietnam War: as with alcohol addiction, the symptoms of PTSD were significantly attenuated.[98] More recently neurofeedback has been used to treat attention deficit disorders: neurofeedback significantly improved results in standardized intelligence tests and the broad clinical picture.[99]

Each year the number of scientific publications on this technology and its clinical applications increases. In June 2011 some 6,000 mentioned neurofeedback. Insurance companies are beginning to cover the costs of therapeutic neurofeedback in the treatment of psychological and psychiatric disorders. Though initiated in the United States, it is used more and more in other countries. Some 6,000 practitioners (physicians, neurologists, or psychiatrists) throughout the world use neurofeedback. Its applications are multiple and its methods of use very versatile. Its recreational aspect is even beginning to enter our homes. The first video games directly controlled by the player's brain waves are on the market, and the sector is expanding. Major companies are investing massively. Young consumers, probably less nervous than their elders, can now play games that affect their mental activity. Progress of neurofeedback seems inevitable: it is poised to invade our daily life and everyday electronic devices, such as computers and mobile telephones. We must prepare to enter a world in which mental hygiene will become the norm, like dental hygiene a century ago, and which is not far away.[100]

The Transhuman Brain

Transhumanism is widely established in the United States and northern Europe. It is not a sect but a philosophy that encourages the improvement of mankind through convergent technology represented by nanotechnology, biotechnology, information technology, and cognitive science (NBIC). The last mentioned is at the heart of the transhumanist project and is officially encouraged by research institutions. The adepts of transhumanism aim to surpass man, which they consider an imperfect species, by enhancing humanity. Transhumanists dream of immortality for a posthuman creature largely produced by human genius and endowed with physical and especially intellectual capacities beyond what modern man can conceive. Man would no longer be a creature but would be his own creator. The future of humanity would be radically transformed by technology. We can envisage man being modified, in terms of youth, increased intelligence through biological or artificial means, the ability to modulate his own psychological state, and the abolition of suffering. These projects are in line with some long-term projects in brain research.[101] For transhumanists the convergence of the four NBIC technologies should enhance human intellectual and physical performance but also permit communication between individuals through the interconnection of their brains to create a true "collective consciousness."

EPILOGUE

I would therein describe men, if need be, as monsters occupying
a place in Time infinitely more important than the restricted one
reserved for them in space, a place, on the contrary, prolonged
immeasurably since, simultaneously touching widely separated
years and the distant periods they have lived through—between
which so many days have ranged themselves—they stand like
giants immersed in Time.

—MARCEL PROUST, *TIME REGAINED*
(TRANS. STEPHEN HUDSON)

TO FEEL GOOD IS AN IDEAL offered by one's brain, as it conducts our
thoughts and actions according to the body's rhythm in response to the
solicitations of the world. True happiness is restrained and far from soul-
destroying excesses. A custom-made brain is one made to the measure of
man. It is the heritage of all, to be shared by all; it is unique yet societal,
unable to exist without the presence of others.

Neuroscience brings a host of data bearing witness to the changeable
and proteiform character of our brain, its dynamism and its unstable
equilibrium. Far from being immutable, the brain is flexible and enjoys
a phenomenal capacity for adaptation. As we learn, new nerve cells are
produced, and new connections are established or reinforced while others are eliminated. This ability of the brain to reconfigure itself keeps it
alive, reactive, and able to solve problems. It is precisely this plasticity
that allows man to escape biological determinism and conformism and
offers him freedom to create and imagine, which distinguishes *Homo
sapiens* from his more or less distant cousins. Thanks to this amazing machine, whose secrets we are beginning to pierce, man remains the only
animal able to escape the dictatorship of genes and hormones.

Analysis of the human genome shows that it is not much more complex than that of a mouse or a fly. However, we have acquired unique mental functions that enable us to express our empathy toward others. We owe these faculties in part to our prefrontal cortex and the freedom it procures. However, this quest for freedom began very early: we find it in fish, and so we need to remain humble as we face life. This book is intended as a homage to not only the brain but also to individuation, which is never quite completed during our all too brief existence.

NOTES

1. Introduction

1. Nicolas Steno was born in Copenhagen into a Protestant family, and he died in Schwerin, Germany. He studied humanities and science at university, then medicine, first in his home town and ultimately in Leiden. Later he converted to Catholicism and published several papers on comparative anatomy. We owe to him the differentiation between white and gray matter in the brain. With views close to those of Baruch Spinoza (1632–1677), Steno estimated that knowledge of the rules that governed natural law originated in the analysis of existing theories, in an experimental approach in which visual observation was crucial, and in conclusions drawn from experimental data so obtained. He was beatified by Pope John Paul II in 1988 for his passionate and insatiable quest for scientific and theological truth.

2. In his *Discours sur l'anatomie du cerveau* (Treatise on the anatomy of the brain) published in Paris in 1669, Nicolas Steno gathered and synthesized the essentials of contemporary knowledge. It was a truly scientific and philosophical undertaking in which Steno accused Willis and Descartes of perverting data on the shape of the brain to fit their speculation about the seat of the soul.

3. Readers wishing to broaden their knowledge of the basis of modern neuroscience can consult G. M. Shepherd, *Creating Modern Neuroscience: The Revolutionary 1950s* (Oxford: Oxford University Press, 2010), among others.

4. We here refer to today's trend for neuroscience to be shouldered by medical imaging.

5. C. P. E. Zollikofer et al., "Virtual Cranial Reconstruction of *Sahelanthropus tchadensis*," *Nature* 434 (2005): 755–759.

6. The first stage of the emergence of modern man can be attributed to the invention of bipedalism more than seven million years ago, when a simian began

to move while standing vertically. In this new posture vision became primary while the nose was moved away from bad odors.

7. The second stage of the origins of humanity was marked by the multiplication of bipedal species according to the process that anthropologists call *adaptive radiation*.

8. The Omo valley is famous worldwide because of the discovery of the remains of a wide variety of hominins that inhabited it over several millennia. Because this region bears witness to important stages in man's cultural development, it was awarded World Heritage Status by UNESCO in 1996.

9. The third revolution at the origin of modern man was the considerable enlargement of the volume of his brain at the time of the emergence of the genus *Homo* (*erectus* then *habilis* then *sapiens*). The fourth and last stage was the acquisition of articulate language, with the development of consciousness of self, artistic imagination, and new aptitudes for technological innovation.

10. Y. Coppens, *L'histoire de l'homme* (Paris: Odile Jacob, 2008).

11. The scarcity of finds, some even unique, explains the extreme fragility of modern hypotheses in the fields of anthropology and paleontology.

12. With the discovery of the *FOXP2* gene, associated with articulate language, anthropologists formulated the hypothesis of a possible hybridization between the two species because *FOXP2* arrived in the *H. sapiens* genome through contact with Neanderthals.

13. So the two species were not completely reproductively isolated, which violates the very definition of "species." It seems that Neanderthals and *H. sapiens* "met" in the Middle East about 80,000 years ago before the colonization of Europe some 40,000 years ago. This meeting took place at the time of their coming out of Africa, when *Homo erectus* decided to emigrate to the Old World. This cross fertilization is seen today by a stronger Neanderthal influence in Europeans than in Papuans. For more information, see R. E. Green et al., "A Draft Sequence of the Neanderthal Genome," *Science* 328 (2010): 710–722.

14. D. Reich et al., "Genetic History of an Archaic Hominin Group from Denisova Cave in Siberia," *Nature* 468 (2010): 1053–1060.

15. According to Kant, man is born twice: first, the day he is born to life and, second, the day he is born to culture.

16. Primary altriciality refers to our being born without immediate competence.

17. This geological period began some 200 million years ago and ended about 140 million years ago. It was marked by the apogee of the dinosaurs and the appearance of *Archaeopteryx*.

18. Surprisingly, the modern concept of cerebral plasticity, a major topic in neurobiology, dates from more than a century ago. Santiago Ramón y Cajal (Nobel laureate in 1906), the founder of modern neuroscience, with his neuronal theory of 1888, proposed in his *Histologie du système nerveux de l'homme et des vertébrés* (1911) an explanation of the phenomena of late learning in the adult by modification of the contacts between neurons.

19. Modern considerations of cerebral handicap challenge the notion of critical period or at least put its unassailability into perspective.

20. In developmental biology, *neoteny* refers to the possibility of retaining juvenile features in adults of the same species, or even the possibility of attaining sexual maturity by an organism still at the larval stage. The first instance is illustrated by the human brain, which reaches its final form very slowly. The implication is extreme vulnerability of young humans, together with long dependence on adults, for socialization is an inevitably long and energy-consuming stage in the formation of viable and autonomous individuals.

21. The term *plasticity* here means a property allowing a change in form or function of a cell, a circuit, or an organ.

22. Donald Hebb (1904–1985) was a Canadian psychologist who developed a theory of learning. His work involving models of neuronal networks was decisive for the evolution of cognitive science and artificial intelligence. In his famous work *The Organization of Behavior* (1949), he proposed a simple rule involving modification of the efficiency of contacts between nerve cells depending on the degree of activity in the neurons they linked. Thus he launched a cellular basis for learning that still inspires numerous experimental and theoretical studies in neurology and psychiatry.

23. M. Foster and C. S. Sherrington, *A Text Book of Physiology*, part 3: *The Central Nervous System* (New York: Macmillan, 1897). According to the principle we call synaptic plasticity, neural circuits are constantly subjected to regulation by which the efficiency of contacts between nerve cells is adjusted depending on the overall activity of the circuit. Thanks to this principle, neurons can be recruited to participate in the global activity of a synchronous group of neurons.

24. J. P. Changeux, P. Courrège, and A. Danchin, "A Theory of the Epigenesis of Neural Networks by Selective Stabilization of Synapses," *Proceedings of the National Academy of Sciences* 70 (1973): 2974–2978.

25. The Swiss psychologist proposed a general theory of the genesis of knowledge centered on the faculties of adaptation from the newborn to the young adult. This acquisition of knowledge was supposed to occur as a series of steps that, once passed, were irreversible.

26. F. Chollet et al., "Fluoxetine for Motor Recovery After Acute Ischaemic Stroke (FLAME): A Randomised Placebo-Controlled Trial," *Lancet Neurology* 10 (2011): 123–130.

27. I. Slutsky et al., "Enhancement of Learning and Memory by Elevating Brain Magnesium," *Neuron* 65 (2010): 165–177. Magnesium is present in mineral water and is a basic ingredient in numerous foodstuffs, notably chocolate, but also wholegrain cereals, beans and lentils, dried fruit, and cashews.

28. For more information on the extraordinary properties of stem cells, see N. M. Le Douarin, *Les cellules souches, porteuses d'immortalité* (Paris: Odile Jacob, 2007).

29. For the reader wishing to read about this in depth, see Ray Kurzweil, *The Singularity Is Near* (New York: Viking Penguin, 2005).

2. And Then There Was Shape

1. The embryonic stage produced by holoblastic cleavage of an ovum, that is to say when the segmentation divides the cytoplasm completely. *Morula* owes its origin to the fertile imagination of scientists who saw in their microscope a cell mass that resembled a mulberry (*Morus*).

2. The Gestalt psychological theory fits this philosophy. Its inspiration lies in some of Goethe's basic concepts. In the nineteenth and twentieth centuries Christian von Ehrenfels and his colleagues developed a theory according to which a complex percept is never a simple juxtaposition of elementary percepts. It postulates that we never perceive single elements but always an ensemble, a whole, a shape, reflecting relationships between elements.

3. We propose to proceed according to a scientific process borrowed from comparative anatomy. It supposes that the morphological structure of organisms obeys strict rules and that the shape of organs is closely related to their function. This postulate is essentially based on the criterion of *homology* as defined by Richard Owen in 1843, comparing structures with the same organizational plan but of which the anatomy differed because they served different functions. In other words, homology was a structural resemblance as exemplified by the comparison of the skeleton of a human hand and a bat's wing. This similarity constitutes a classic example of the anatomical criteria of homology. On the other hand, *analogy* was a similarity in function attributable to convergent adaptation to similar lifestyles. The comparison of the fin of a fish with that of a whale illustrates functional but not structural similarity.

4. It is important to note that the notion of genes participating in the development of the embryo links not only physiology to development but also development to evolution and ecology.

5. *Encephalon* is a term from the Greek for "in the head" (*en, kephalos*). It is the most anterior part of the central nervous system within the skull. It consists of three structures, the cerebral hemispheres, the cerebellum, and the brainstem, which sit on top of the spinal cord. The encephalon controls most of the functions of the nervous system.

6. W. J. Gehring, "The Homeobox in Perspective," *Trends in Biochemical Sciences* 17 (1992): 277–280; T. Montavon and D. Duboule, "Landscapes and Archipelagos: Spatial Organization of Gene Regulation in Vertebrates," *Trends in Cell Biology* 22 (2012): 347–354.

7. Thomas Hunt Morgan (1866–1945) was a pioneer of genetics. A professor of experimental zoology at Columbia University (1904), he later moved to the California Institute of Technology at Pasadena (1928), where he remained for the rest of his career. He was an expert in zoology and studied phenotypic (the observable features of an organism) variations in the fruit fly (the famous *Drosophila*). His observations enabled him to discover the crucial role of chromosomes in heredity. His major contributions to genetics earned him the Nobel Prize in Physiology or Medicine in 1933. We owe to him the adoption of *Drosophila* as a model in modern biology.

8. Etienne Geoffroy Saint-Hilaire was a French naturalist, born at Etampes in 1772, who died in Paris in 1844. He was mainly interested in anatomical research, notably the bony skeleton, to which he gave precedence over the nervous system. We owe to him the foundation of the menagerie at the Museum of Natural History in Paris. "It seems that nature is confined within certain limits and has formed all living creatures according to a single plan, essentially the same in principle, but which she has varied in a thousand ways in all accessory parts." E. Geoffroy Saint-Hilaire, "Mémoire sur les rapports naturels des Makis Lémur," *Magasin Encyclopédique* 1 (1796): 20–50. The literature on Saint-Hilaire is abundant, and we cannot attempt to provide an exhaustive bibliography here. For a chronology of his work the reader can consult J. L. Fischer, "Chronologie sommaire de la vie et des travaux d'Etienne Geoffroy Saint-Hilaire," *Revue d'Histoire des Sciences* 25 (1972): 293–300. The famous controversy between Cuvier and Geoffroy Saint-Hilaire of 1830 concerned the conclusions of an article submitted to the Academy of Sciences that proposed the fusion of vertebrates and insects at the branching off of the mollusks by maintaining that there was only one system of organic composition. E. Geoffroy Saint-Hilaire, *Principes de philosophie zoologique, discutés en mars 1830 au sein de l'Académie royale des sciences* (Paris: Pichon et Didier, 1830). For a comprehensive history of the concept of organizational plan, see H. Le Guyader, "Le concept de plan d'organisation: quelques aspects de son histoire," *Revue d'Histoire des Sciences* 53 (2000): 339–379.

9. The cephalochordates are on the border between invertebrates and vertebrates, so they provide a model for understanding the origin of vertebrates. Philippe Vernier and his team demonstrated the consistency of neurotransmitter systems implicated in desire and affective processes of punishment and reward, notably in the older species of the vertebrate branch.

10. The term *craniates*, used today for vertebrates, shows the importance of their new head: Carl Gans and Glenn Northcutt first emphasized its fundamental role in the emergence of the new branch. C. Gans and R. G. Northcutt, "Neural Crest and the Origin of Vertebrates: A New Head," *Science* 220 (1983): 268–274.

11. The work of Nicole Le Douarin has demonstrated the important role of this transitory embryonic structure, the neural crest, in the formation of the nervous system of vertebrates and the development of the head. For more details see N. M. Le Douarin and C. Kalcheim, *The Neural Crest*, 2nd ed. (Cambridge: Cambridge University Press, 1999).

12. A. Joliot et al., "Antennapedia Homeobox Peptide Regulates Neural Morphogenesis," *Proceedings of the National Academy of Sciences* 88 (1991): 1864–1868. Some complementary information concerning this family of *Hox* genes: They are involved in cellular identity along the anteroposterior axis of vertebrates, like homeotic genes in insects. The expression of *Hox* genes has been studied in axial embryonic organs of vertebrates, such as the neural tube or the vertebral column. In vertebrates, we recognize four homologous complexes of *Hox* genes carried on different chromosomes. These genes are *paralogous*; they are derived from a succession of two duplications carried on a single complex of a common ancestor having probably existed before the divergence of insects and vertebrates. It is the

accumulation of successive mutations and duplications during evolution from an initially single homeotic complex that has led vertebrates to develop under the influence of four *Hox* complexes. To better understand the *Pax* gene family, consult R. Wehr and P. Gruss, "*Pax* and Vertebrate Development," *International Journal of Developmental Biology* 40 (1996): 369–377.

13. Genes that control cell position during embryogenesis and organogenesis are called *selector genes*.

14. Group of organisms sharing the same ancestor and expressing certain common features. Based on the work of the Swedish naturalist Carl Linnaeus (1707–1778), the hierarchy of taxons in the animal world is: kingdom, phylum, class, order, family, genus, species.

15. Derived from the Greek *meta* for "after" and *zoon* for "animal," this is a modern name for the taxon composed of multicellular animals, as opposed to protozoans, the single-celled organisms.

16. The phylotypic stage is a period of strong resemblance among many embryos, even if their adult forms are very different. This stage is controlled by a family of so-called homeobox genes. See J. M. W. Slack et al., "The Zootype and the Phylotypic Stage," *Nature* 361 (1993): 490–492.

17. A few definitions might be useful here in understanding the complex jargon of embryologists. The homeotic gene is characterized by a nucleotide sequence common to all homeotic genes: the homeobox. The homeotic gene codes for homeoprotein, a transcription factor. It has an amino acid sequence common to all homeoproteins, called the homeodomain. Homeobox is a sequence of 180 nucleotide base pairs that code for the homeodomain, itself corresponding to a sequence of sixty amino acids, of which the three-dimensional conformation specifically recognizes regulatory regions of certain genes. But not all homeobox genes are homeotic genes.

18. The homeotic effect of a mutation is always polarized. In the case of the *Antennapedia* mutation, we find legs in the place of antennae. Homeotic transformation takes place from back to front if a *Hox* gene loses its function and from front to back when the mutation leads to an increase in function. This polarized effect of the mutations of *Hox* genes is a general rule applicable from insects to vertebrates.

19. Homeotic genes play a fundamental and universal role in the formation of the embryo. They inform cells of their position along three possible axes: anteroposterior, dorsoventral, and mediolateral (left-right). These genes act at the same place, whatever the organism, as topological markers common to all metazoans. The processes of embryological development, so different in a fly or a mouse, and more generally the diversity of the animal kingdom, depend on the participation of similar genes (duplicated or mutated) belonging to the same great family of genes.

20. The characteristic of the nervous system of the arthropod is that it takes the form of ventrally situated ganglia. All arthropods possess a pair of cerebral ganglia and another pair of subesophageal ganglia, but the anterior part of this nervous system is not particularly developed. Like a rope ladder, each pair is linked by a commissure, and successive pairs are linked by connectives. Only some higher arthropods have ganglia that fuse into one or several large nerve clusters, but we are

still a long way from the outline of a head. Ventrally situated central nervous systems are called hyponeurian; dorsally situated central nervous systems, epineurian.

21. This family of genes is inherited from a common ancestor dating from the beginning of the Cambrian era, that is to say some 550 million years ago.

22. E. B. Lewis, "A Gene Complex Controlling Segmentation in *Drosophila*," *Nature* 276 (1978): 565–570.

23. Amphioxus, the lancelet, is the best-known representative of the cephalochordates. The name is from the Greek *amphis* ("both") and *oxys* ("sharp"). Both extremities of the body of this ancestor of the vertebrates are "sharp," so it has no head. It looks like a sort of immobile worm buried in the seabed, feeding by filtering plankton stirred up by waves. Its position in the animal world is remarkable because this tiny creature seems almost unchanged for around 520 million years. Although officially an invertebrate, amphioxus shares common features with vertebrates such as, for example, a hollow nerve cord along its back like the vertebrate spinal cord. Equally, the amphioxus notochord, a rigid but supple cartilaginous structure that supports its body, strangely resembles a vertebral column. Because of its position in the animal kingdom, amphioxus is a key to understanding the origin of vertebrates—the animal group with bony skeletons that includes fish, amphibians, reptiles, birds, and mammals. Like an ancient creature sealed inside a block of amber, amphioxus belongs to that exclusive group of "living fossils," as Darwin called creatures alive today but remarkably similar to their fossil ancestors, and it is thus of great scientific value.

24. The chordates form one of the major metazoan phyla, and it is estimated that there are 51,000 living species. The cephalochordates are considered one of the closest relatives of vertebrates within the chordates, which explains why amphioxus remains easily the most-used model to understand the rules of evolutionary development, from which stems the term "evo-devo," often employed by specialists.

25. In 1995, the Nobel Prize in Physiology or Medicine was awarded to the scientists who discovered the existence of homeotic mutations. They were Edward Lewis, Christiane Nüsslein-Volhard, and Eric Wieschaus, just a century and a year after the formulation of the concept of homeosis by William Bateson (1861–1926).

26. J. Deutsch and H. Le Guyader, "Le zootype neuronal," *Journal de la Société de Biologie* 194 (2000): 71–79.

27. This group consists of about thirty phyla, among which number the arthropods, nematodes, flatworms, and vertebrates.

28. There exist eight homeotic genes in *Drosophila*, among which *Antennapedia* and *Ultrabithorax* have been studied most. In vertebrates (for example, reptiles, mice, and man), there are four homeotic complexes. The most posterior genes have been duplicated several times, and so we now recognize thirty-eight human homeotic genes. Homeobox genes code for proteins that have a homeodomain. They are transcription factors of the helix-loop-helix family. Homebox genes involve numerous living organisms (plants, animals, fungi), but homeotic genes of the *Hox* family are specific to animals.

29. D. Duboule and G. Morata, "Colinearity and Functional Hierarchy Among Genes of the Homeotic Complexes," *Trends in Genetics* 10 (1994): 358–364.

30. "I us the term 'hopeful monster' to express the idea that mutants producing monstrosities may have played a considerable role in macroevolution. A monstrosity appearing in a single genetic step might permit the occupation of a new environmental niche and thus produce a new type in one step. A Manx cat with a hereditary concrescence of the tail vertebrae, or a comparable mouse or rat mutant, is just a monster. But a mutant of Archaeopteryx producing the same monstrosity was a hopeful monster because the resulting fanlike arrangement of the tail feathers was a great improvement in the mechanics of flying." R. B. Goldschmidt, *The Material Basis of Evolution* (New Haven, Conn.: Yale University Press, 1940).

31. The only surviving cephalochordate today is the amphioxus. The Cambrian period of the Paleozoic era began 540 million years ago. The fish were the first animals to possess an internal skeleton. Before them, animals were equipped with an external skeleton, like a shell. The first land animals appeared about 400 million years ago.

32. Deutsch and Le Guyader, "Le zootype neuronal."

33. The major genetic innovations depend on the duplication of a gene and then a random accumulation of mutations. Gene mutations either can inactivate it (a pseudogene) or bring new functions to the organism, beneficial or harmful. Indeed, "tinkering" with life cannot predict the definitive outcome.

34. See R. Thom, *Structural Stability and Morphogenesis: An Outline of a General Theory of Models*, trans. D. H. Fowler (Reading: W. A. Benjamin, 1975). The following passage reflects his fundamental theory: "The concept of structural stability seems to me to be a key idea in the interpretation of phenomena of all branches of science (except perhaps quantum mechanics) for reasons to be given later. Meanwhile we note only that forms that are subjectively identifiable and are represented in our language by a substantive are necessarily structurally stable forms." See also Ilya Prigogine, *The End of Certainty* (New York: The Free Press, 1997): "We now recognize that equilibrium physics gave us a false image of matter. . . . Once again, we are faced with the fact that matter at equilibrium is 'blind,' while in nonequilibrium it begins to 'see.'" Benoit Mandelbrot (1924–2010) was a brilliant mathematician. See his *The Fractal Geometry of Nature* (San Francisco: W. H. Freeman, 1982). See also D. Ruelle and F. Takens, "On the Nature of Turbulence," *Communications in Mathematical Physics* 20 (1971): 167–192; 23 (1971): 343–344; and S. Newhouse, D. Ruelle, and F. Takens, "Occurrence of Strange Axiom A Attractors near Quasi Periodic Flows on Tm, m ≥ 3," *Communications in Mathematical Physics* 64 (1978): 35.

35. D. W. Thompson, *On Growth and Form* (Cambridge: Cambridge University Press, 1917). Thompson explored how life is subject to physical constraints. Among other things, he explained how the world is exposed to physical forces that biologists neglect, preferring to explain everything by the rag-bag law of natural selection. Inspired by philosophers such as Francis Bacon, Thompson refused to accept natural science as a discipline seeking finalistic explanations.

36. "The neuronal zootype is a particular group of genes of which the primordial, primary, and primitive function is to determine neuronal pathways that harmonize

the central nervous system and the body plan." Deutsch and Le Guyader, "Le zootype neuronal."

37. Metazoans: we can distinguish parazoans, mesozoans, and eumetazoans. The cells of parazoans and mesozoans form clusters. Those of eumetazoans undergo true differentiation. These organisms have embryonic germ layers capable of forming different tissues and hence organs.

38. For in-depth information, the reader should consult the works of Alain Prochiantz.

39. We owe the concept of the three embryonic germ layers to the German embryologist Karl von Baer (1792–1876).

40. There is a more complete analysis of the development of the embryo in Le Douarin and Kalcheim, *The Neural Crest*.

41. In nature, three different strategies exist for building an embryo: First, like *Caenorhabditis elegans*, a small, transparent roundworm that only contains 959 somatic cells and 302 neurons. In its case, the segmentation of the egg is called mosaic. After the first division of the egg into two cells, one takes up a position at the anterior pole, the other posterior. If one cell is lost, no descendants are possible. Its fate is sealed at its first division. Second, like insects, in which development is syncytial; that is to say, there is no segmentation. They form two axes: one depends on bicoid and nanos genes, the other is dorsal. Third, like vertebrates: the egg segments, and if some cells are lost, the remaining cells are able to compensate. This property illustrates the absence of a predefined cell lineage: the cells' fates will only be fixed later, during gastrulation.

42. The stage when the different members of a phylum most closely resemble the others is the phylotypic stage that we described earlier.

43. Ernst Haeckel (1834–1919) suggested that "the series of forms by which the individual organism passes from the primordial cell to its full development is only a repetition in miniature of the long series of transformations undergone by the ancestors of the same organism from the most remote times to our own day."

44. This is a recent multidisciplinary approach in which embryologists collaborate with molecular biologists, evolutionists, paleontologists, morphologists, physiologists, zoologists, and bioinformaticians. Two works summarize this discipline: N. Shubin, *Your Inner Fish: A Journey Into the 3.5-Billion-Year History of the Human Body* (New York: Pantheon, 2008); and S. B. Carroll, *Endless Forms Most Beautiful: the New Science of Evo Devo and the Making of the Animal Kingdom* (New York: Norton, 2005).

45. For Spemann, an organizer was a region of the embryo capable of inducing the formation of new tissues and organizing them in space. See C. Waddington, *Organisers and Genes* (Cambridge: Cambridge University Press, 1940).

46. For readers wishing to learn more about the era of the discovery of neural induction and its consequences for contemporary ideas about animal embryology, we recommend Le Douarin and Kalcheim, *The Neural Crest*.

47. A substance that plays a crucial role in the formation of a pattern and of which the concentration varies in space and time is called a morphogen.

48. Le Douarin and Kalcheim, *The Neural Crest.*

49. The same intuition prompted Jean-Baptiste Lamarck, the father of the theory of transformism, to distinguish vertebrates from invertebrates. In his opening speech on the 21st day of Floreal, Year VIII (May 11, 1800), he clearly pronounced the basis of this distinction: "For several years I have pointed out in my Lectures at the Museum that a consideration of the presence or absence of a vertebral column in the body of animals separates the animal kingdom into two very different categories that one can in a way consider as two great families of the first order. . . . All known animals can therefore be distinguished quite remarkably: 1. animals with vertebrae. 2. animals without vertebrae." J. B. Lamarck, *Système des animaux sans vertèbres* (Paris: Déterville, 1801).

50. The essential characteristic of vertebrates is the superimposition of four organs: a nervous system, a dorsal neural tube, an aorta, and a digestive tract.

51. For more than two centuries, biologists have debated the origin of vertebrates. In particular, the history of their classification is marked by scientific quarrels. It began in 1758 with Carl Linnaeus, who distinguished four classes: fish, amphibians (including reptiles, frogs and toads, and certain fish), birds, and mammals. In 1816, Henri Marie Ducrotay de Blainville suggested that reptiles be considered as a fifth class distinct from the amphibians. Then in 1926 the Swedish paleontologist Erick Helge Osvald Stensiö proposed that a distinction be made between agnathan (jawless) fish and gnathostomes (with jaws). Today, the fish can be divided into two classes: Chondrichthyes (cartilaginous fish, from the Greek *chondros*, "cartilage," and *ichthys*, "fish") and Osteichthyes (bony fish), and the vertebrate subphylum contains seven classes: jawless fish, cartilaginous fish, bony fish, amphibians, reptiles, birds, and mammals.

52. All other vertebrates belong to the infraphylum of gnathostomes, the jawed vertebrates. Once very abundant on our planet, the agnathans are today only represented by two species: hagfish and lampreys.

53. In vertebrates, cephalization (the tendency for nerve centers to be located anteriorly) increases from older groups, such as fish, to more recent ones, like mammals. But in invertebrates the process does not obey the same phylogenetic rules. In the transition from invertebrates to vertebrates, the general form of the body, the diversity of sensory organs, the way of life, and the variety of behavior are more important as evolutionary forces for cephalization.

54. This phylum is characterized by the possession of a rigid skeletal axis called the notochord. Chordates share a common ancestor with the prochordates, represented by the cephalochordates (like amphioxus) and the urochordates (like the ascidians, the sea squirts). This divergence is thought to have occurred more than 750 million years ago.

55. The telencephalon is part of the embryonic prosencephalon, or forebrain. It will form the mature cerebrum.

56. Adding to the theory of Karl von Baer, according to which the mesoderm produces the mesenchyme and the bones of the skeleton, we now know that part of the ectoderm also participates in the formation of skeletal tissue.

57. A single embryonic vesicle of which the caudal extension produces the diencephalon during the formation of the central nervous system of vertebrates.

58. Classically we recognize three types of glia: astrocytes, microglia, and oligodendrocytes. They occupy the spaces around the neurons, with which they have a common embryonic origin. They are not excitable, but neurobiologists regularly discover new functional properties. For a long time it was thought that their main functions were in the nutrition of neurons and in increasing the speed of impulse conduction by insulating nerve fibers with myelin. Over the last few decades, their status has been upgraded. No longer merely supporting cells, glia are now seen to fulfill roles as varied as regulation of the cerebral vascular system, active participation in transmission between neurons, and playing the part of stem cells in adults.

59. The genes involved in genetic control of embryonic development interact directly with the DNA molecule to regulate the expression of other genes. They belong to the very large family of transcription factors, and as they exert a selective function they are also called selector genes.

60. The term "epigenetic" is attributed to Conrad Waddington (1905–1975) in 1942. He referred to "the branch of biology which studies the causal interactions between genes and their products which bring the phenotype into being." But the origin of the concept goes back to Aristotle (384–322 bce), for whom the development of an organic form was derived from the unformed. Epigenesis refers to the overall influences exerted by the cellular or physiological environment on gene expression in a cell. In neurobiology, epigenesis signifies that even if there exists a preformed program, its expression is variable. Interaction of an individual with the environment is firmly inscribed in the physiology of its neurons, and the term applies also to the effects of neural activity on the form or function of circuits and neurons.

61. Programmed cell death, also called apoptosis or "cell suicide," plays an important role in outlining the contours of an organ during morphogenesis and, later, in the function of adult tissues. For example, without programmed cell death interdigital membranes would not be broken down, and the fetus would have webbed hands and feet.

62. Numerous very complex molecules, often polymers, regulate the functional activity of the extracellular matrix, which consists of two parts, the interstitial matrix and the basement membrane, a dense fibrous sheet. Recent work emphasizes the importance of interactions between nerve cells and the extracellular matrix in the control of cellular processes related to neuronal development: proliferation, migration, and differentiation. More details can be found in Michel Imbert, *Traité du cerveau* (Paris: Odile Jacob, 2006).

63. Information between a neuron and its target is exchanged at the synapse, of which there are two types: chemical, where neurotransmitters relay the nerve impulse; and electrical, which allow the passage of the impulse without a chemical intermediary.

64. For more information, see Deutsch and Le Guyader, "Le zootype neuronal."

65. S. Sugiyama et al., "Experience-Dependent Transfer of Otx2 Homeoprotein Into the Visual Cortex Activates Postnatal Plasticity," *Cell* 134 (2008): 508–520; S. Sugiyama, A. Prochiantz, and T. K. Hensch, "From Brain Formation to Plasticity: Insights on Otx2 Homeoprotein," *Development, Growth & Differentiation* 51 (2009): 369–377.

66. E. Téglás et al., "Pure Reasoning in 12-Month-Old Infants as Probabilistic Inference," *Science* 332 (2011): 1054–1059.

3. The Masterpiece

1. The allocortex is formed quite early, between the eighth and twelfth weeks of gestation. It has from three to six layers depending upon the region. The mesocortex is a transitional form between neo- and allocortex, found in the adult parahippocampal and cingulate cortex. The allocortex also contains the archicortex, the olfactory cortex of the rhinencephalon. It is found in the adult dentate gyrus and hippocampus. It controls basic behavior needed for the survival of the species. The paleocortex (also part of the allocortex) is also linked to the olfactory system, in the adult olfactory bulbs and tubercles, and the pyriform and entorhinal cortex. It controls motivation, selective attention, and emotive reactions. It assists in selection of behavior as a function of a repertory acquired during earlier learning.

The neocortex is formed between the twelfth and twenty-eighth weeks of gestation. Whatever the region, it has six cell layers. The size of the neocortex varies from one species to another. Fish and amphibians have none, in the shrew the neocortex is 20 percent of the total brain weight, and in man 80 percent. R. Nieuwenhuys, H. J. Ten Donkelaar, and C. Nicholson, *The Central Nervous System of Vertebrates*, vol. 3 (Berlin: Springer, 1998). In fact, it is during the transition from nonhuman to human primates that the neocortex develops most. Among all the neocortical regions the prefrontal cortex has undergone the greatest expansion in man.

2. See, for an example of such a tour, J. D. Vincent, *Voyage extraordinaire au centre du cerveau* (Paris: Odile Jacob, 2007).

3. The peripheral nervous system is a prolongation of the central nervous system (CNS) and consists of those nerve cells and fibers outside the brain and spinal cord. These nerves are attached to the CNS and ramify as fine branches innervating the whole body. There are cranial nerves (ten pairs from the brain) and spinal nerves (thirty-one pairs from the spinal cord).

4. There exists a school of thought, computationism, according to which mental activity can be summed up as a logical series of operations executed very quickly. This hypothesis was first defended in 1943 by Warren McCulloch and Walter Pitts, for whom thought could be seen as a series of symbolic representations

working like a machine or a computer. W. S. McCulloch and W. H. Pitts, "A Logical Calculus of the Ideas Immanent in Nervous Activity," *Bulletin of Mathematical Biophysics* 5 (1943): 115–133.

5. The brain does not exercise a single function on which evolution could act. It is rather a collection of systems, which theoreticians call modules, controlling various cerebral functions. We should note that evolution acts on individual modules rather than on the whole brain. Although evolution favored an overall increase in brain size, for example in primates, its influence is felt most in specific systems. In man, for example, the result is a relatively small auditory area compared with that of a cat, whereas the human visual area developed much more.

6. One has for long counted some ten glia for each neuron, but for some years this ratio has been under revision. For example, Frederico Azevedo and his colleagues estimated that there exist eighty-six billion neurons and eighty-five billion glial cells in the adult human brain. F. A. C. Azevedo et al., "Equal Numbers of Neuronal and Nonneuronal Cells Make the Human Brain an Isometrically Scaled-up Primate Brain," *Journal of Comparative Neurology* 513 (2009): 532–541. This ratio varies in different cerebral structures, being greater in subcortical regions but never exceeding 2:1 in favor of glia in the cortex. Over the whole human brain the ratio is almost 1:1, similar to the figure for other primates and mammals.

7. In a chemical synapse between two neurons, the neuron from which the nerve impulse is coming is termed presynaptic. The one that receives the chemical neurotransmitter is postsynaptic.

8. The genotype is the ensemble of genes carried on the chromosomes of a given cell. The phenotype, by contrast, is the ensemble of an individual's or organ's traits, dependent on the expression of its genes and their interaction with the environment.

9. As Alain Prochiantz emphasized, "the concept of an individual is not the same for all species. It is not the same for a worm that has nothing, or almost nothing, to distinguish it from its neighbor, as for a vertebrate in which the structure of the nervous system carries the material trace of its individual history." A. Prochiantz, *Les anatomies de la pensée: à quoi pensent les calamars?* (Paris: Odile Jacob, 1997).

10. This backward tipping of the head enabled the acquisition of an upright stance (bipedalism). Incidentally, it facilitated the descent of the larynx (which remains higher in the great apes), which allows vocalization.

11. The frontal lobe is limited posteriorly by the central sulcus (of Rolando), inferiorly by the lateral fissure (of Sylvius), and on the medial surface by the cingulate sulcus. The parietal lobe extends over the superior and middle parts of the lateral surface of the hemisphere. It is limited anteriorly by the central sulcus, inferiorly by the lateral fissure, and posteriorly by the parieto-occipital sulcus. The occipital lobe forms the posterior pole of the hemisphere, and the temporal lobe its middle and inferior parts. The insula is in the depths of the lateral fissure: you have to retract the banks of the fissure to see it. The cingulate gyrus surrounds the corpus callosum on the medial aspect of the hemisphere; posteriorly it joins the parahippocampal gyrus of the temporal lobe to form the limbic lobe.

12. At the end of the eighteenth century, the fashion was to search for precise localization of major mental functions—researchers in this area included Vincenzo Malacarne (1744–1816) and Franz Joseph Gall (1758–1828). The latter was the founder of *phrenology*, according to which each mental function was associated with a region of the brain as diagnosed from the outside of the skull. Later, separate from the principles of phrenology, Paul Broca (1824–1880) and Carl Wernicke (1848–1905) established links between certain cortical areas and cerebral functions from observations of clinical cases. At the beginning of the twentieth century Korbinian Brodmann (1868–1918) described some fifty cortical areas on the basis of their cellular organization. This "localization" provoked severe criticism from adepts of the theory of conditioned reflexes, described by Ivan Pavlov (1849–1936) then defended vehemently by the behaviorists, for whom the brain was a black box to be ignored while they concentrated on the study of behavior. See L. J. Garey, *Localisation in the Cerebral Cortex* (London: Smith-Gordon, 1994; 3rd ed., New York: Springer, 2005), translated from K. Brodmann, *Vergleichende Lokalisationslehre der Grosshirnrinde in ihren Prinzipien dargestellt auf Grund des Zellenbaues* (Leipzig: Barth Verlag, 1909).

13. The homunculus was discovered by the neurosurgeon Wilder Penfield, who by stimulating the cortex of patients undergoing neurosurgery was the first to describe the somatotopic organization of human cortex. W. Penfield and E. Boldrey, "Somatic Motor and Sensory Representation in the Cerebral Cortex of Man as Studied by Electrical Stimulation," *Brain* 60 (1937): 339–448; W. Penfield and T. Rasmussen, *The Cerebral Cortex of Man* (New York: MacMillan, 1950). The homunculus has four characteristics: (1) the body is segmented into well-defined territories, (2) spatial relationships between these territories do not reflect the physical reality of the body (for example, the hand is next to the face), (3) the relative importance of each part is not respected (the lips are disproportionate to the thorax), and (4) the maps are more or less dynamic. The last property can be seen by, for example, a greater cortical representation of the fingers in professional violinists than amateurs. M. Lotze et al., "The Musician's Brain: Functional Imaging of Amateurs and Professionals During Performance and Imagery," *NeuroImage* 20 (2003): 1817–1829.

14. According to the theoretician Jerry Fodor, the human brain functions on the basis of small specialized programs that he defined as modules. Columns would represent the anatomical substrate for autonomous and modular processing, which would be different from column to column. J. A. Fodor, *Modularity of Mind: An Essay on Faculty Psychology* (Cambridge, Mass.: MIT Press, 1983). In this context we must also cite the pioneering work of David Hubel and Torsten Wiesel, who discovered columns in the visual cortex of the cat and monkey. D. H. Hubel and T. N. Wiesel, "Receptive Fields, Binocular Interaction, and Functional Architecture in the Cat's Visual Cortex," *Journal of Physiology* 160 (1962): 106–154; "Ferrier Lecture: Functional Architecture of Macaque Monkey Visual Cortex," *Proceedings of the Royal Society*, London, Series B 198 (1977): 1–59. Earlier, Vernon Mountcastle showed the way by studying the somatosensory cortex of the cat. V. B. Mountcastle, "Modality and Topographic Properties of Single

Neurons of Cat's Somatic Sensory Cortex," *Journal of Neurophysiology* 20 (1957): 408–434.

15. R. Llinás, *I of the Vortex: From Neurons to Self* (Cambridge, Mass.: MIT Press, 2001). Vilayanur Ramachandran of the University of California–San Diego considered that consciousness emerged when during hominization some neural structures evolved that sent output from primary sensory areas to the thalamus and then, after processing, back to the cortex, where the information became a "metarepresentation." In other words, instead of producing primary sensory representations, the human brain preferred to create representations of representations, a process that might be a basis for symbolic thought. By a bit of juggling, and in improved form, information about the sensation would thus be more easily managed, particularly for language.

16. Homeostasis describes coordinated and in large part automatic physiological reactions indispensable for maintaining a stable internal environment in the living individual.

17. J. D. Vincent, *The Biology of Emotions* (Oxford: Blackwell, 1990).

18. The origin of "addiction" is the Latin *addictus*, "assigned by decree."

19. J. A. Kauer and R. C. Malenka, "Synaptic Plasticity and Addiction," *Nature Reviews Neuroscience* 8 (2007): 844–858.

20. Vincent, *The Biology of Emotions*.

21. R. L. Solomon and J. D. Corbit, "An Opponent-Process Theory of Motivation. I. Temporal Dynamics of Affect," *Psychological Review* 8 (1974): 119–145.

22. J. D. Vincent, *La chair et le diable* (Paris: Odile Jacob, 1996).

4. The Workshop of the Brain

1. For more information concerning the evolutionary aspects of regeneration, see the review: E. M. Tanaka and P. Ferretti, "Considering the Evolution of Regeneration in the Central Nervous System," *Nature Neuroscience Reviews* 10 (2009): 713–723.

2. The ancient metaphor of the Tree of Life proposes an organization of animal and plant species according to a hierarchy in which the common ancestor of two close species enjoys a lower stature because of a greater degree of perfection in its descendants. At the top of the tree we find our own species, even in recent versions. The term "higher" stems from this hierarchical view. We now know that the emergence of a new species does not always imply the loss of a common ancestor.

3. A. Prochiantz, *Les anatomies de la pensée. A quoi pensent les calamars?* (Paris: Odile Jacob, 1997). The dynamics of embryogenesis plays a crucial role in the expression of the phenotype, for transcription is not synchronized over all genes and all cells. In this way mutants can emerge without involving any modification of their genes. An insect will have two pairs of wings if it is a butterfly or a single pair if it is a fly, depending on when the cells participating in the construction of the wings differentiate. For more on this phenomenon of heterochronology in

development, consult the thesis of Florian Ferrand, University of Quebec, Montreal, June 2008.

4. B. A. Reynolds and S. Weiss, "Generation of Neurons and Astrocytes from Isolated Cells of the Adult Mammalian Central Nervous System," *Science* 255 (1992): 1707–1710.

5. J. Altman, "Are New Neurons Formed in the Brains of Adult Mammals?" *Science* 135 (1962): 1127–1128.

6. The author of the highly successful *The Structure of Scientific Revolutions* (Chicago: University of Chicago Press, 1962).

7. We shall also use the term "secondary" neurogenesis to distinguish that in the adult from that during embryogenesis, which is "primary." The latter is part of ontogenesis and includes the developmental period up to sexual maturity.

8. As we emphasized in chapter 1, critical periods during which learning must take place do not exist, strictly speaking, but rather sensitive periods during which learning is more efficient.

9. Neoteny describes the slowing of the physiological development of an animal before it reaches the adult stage so that it remains blocked at a juvenile stage. It is an aspect of pedimorphism, which is the persistence of infantile or embryonic characters in the adult. A well-known example is the axolotl (*Ambystoma mexicanum*), considered as an independent species until, by inducing metamorphosis by hormonal treatment, it was discovered that it was in fact the larva of a salamander.

10. In 1980, François Jacob, Nobel laureate in Physiology or Medicine, noted that man was certainly programmed, but programmed to learn; this was his way of resolving the persistent dilemma between the innate and the acquired. The tendency to enclose man in biological determinism reduces him to a slave of logic, whereas it is exactly the freedom to invent and imagine that distinguishes him from other species. Thanks to his brain, man remains alone in escaping from laws dictated by genes and hormones.

11. Phenotypic plasticity is defined as the ability of a genotype to show morphological, physiological, or behavioral modifications in response to the demands of the environment.

12. D. R. Euston, A. J. Gruber, and B. L. McNaughton, "The Role of Medial Prefrontal Cortex in Memory and Decision Making," *Neuron* 76 (2012): 1057–1070.

13. P. Glansdorff and I. Prigogine, *Thermodynamic Theory of Structure, Stability, and Fluctuations* (New York: Wiley-Interscience, 1971).

14. J. P. Changeux and A. Danchin, "Selective Stabilization of Developing Synapses as a Mechanism for the Specification of Neuronal Networks," *Nature* 264 (1976): 705–712.

15. D. Purves and J. W. Lichtman, "Geometrical Differences Among Homologous Neurons in Mammals," *Science* 228 (1985): 298–302.

16. The term "information" is used here to denote the ensemble of messages that enable the organism to enquire, consciously or not, about the state of the world in which it lives or its inner self. This terminology diverges from the initial definition proposed by Claude Shannon in 1948, which considered information as

an observable and measurable entity. At the Bell Laboratories Shannon reflected on the way information was reliably transmitted between two systems, that is to say, seeking to minimize the probability of error.

17. Nevertheless, the myth of an alternative reality is a pillar of modern popular culture. We might cite the film *The Matrix*, in which, without realizing, men live in virtual reality created by superintelligent computers. This film asks the question how we can be certain we are not inhabitants of a fictitious world. As it is extremely difficult to answer, the cinema industry has a bright future.

18. See J. D. Vincent and J. M. Amat, *Pour une nouvelle physiologie du goût* (Paris: Odile Jacob, 2000).

19. This link between smell and memory depends, among other things, on the special pathway taken by olfactory messages. They are not relayed in the thalamus, where all other sensory information converges before continuing to specific areas of the cerebral neocortex. Olfactory information passes from the nasal cavities to diffuse circuits for memory and emotion, without relay or representation in the neocortex. Thus our brain is organized so that a smell wakes a diffuse impression focused by a memory.

20. The rhinencephalon is formed by a broad band of cortex on the inside surface of the cerebral hemispheres. It is an ancient part of the brain and has only two cortical layers. Its derivation from the Greek *rhinos*, for "nose," reminds us of its special relationship with smell. But the rhinencephalon also receives auditory, tactile, and visual information, which it processes in a context of memory or affectivity.

21. Conceptually the discovery of adult neurogenesis in the sensory organ met with no major objections, but that in the olfactory bulb provoked disputes and lively debates.

22. Stem cells of the olfactory epithelium have the same characteristics as those in the central nervous system. They are undifferentiated, capable of autorenewal and pluripotent, that is to say, able to replace the various cell types of the epithelium.

23. More than just aging, it is the lack of mental and physical activity associated with boredom which seems deleterious for intellectual performance in senior citizens. The SHARE project (Survey of Health, Ageing, and Retirement in Europe) involved 85,000 individuals from nineteen European countries and confirmed this hypothesis (http://www.share-project.org/). Tests of memory and language were carried out on people aged fifty or over. The results were unequivocal: there was a considerable decline after retirement. Countries in which retirement is early (France, Poland, Austria, Belgium, Italy) had the worst scores, while in the best (Sweden, Switzerland) retirement is later. This is food for thought in the debate over retirement age.

24. Perhaps one day we shall have new medication to stimulate at will the production of specific new neurons.

25. G. Gheusi et al., "Importance of Newly Generated Neurons in the Adult Olfactory Bulb for Odor Discrimination," *Proceedings of the National Academy of Sciences* 97 (2000): 1823–1828.

26. C. Rochefort et al., "Enriched Odor Exposure Increases the Number of Newborn Neurons in the Adult Olfactory Bulb and Improves Odor Memory," *Journal of Neuroscience* 22 (2002): 2679–2689.

27. T. Shingo et al., "Pregnancy-Stimulated Neurogenesis in the Adult Female Forebrain Mediated by Prolactin," *Science* 299 (2003): 117–120.

28. M. Botvinick and J. Cohen, "Rubber Hands 'Feel' Touch That Eyes See," *Nature* 391 (1998): 756.

29. Studies suggest that 5 to 10 percent of the population report, at least once, an out-of-body experience, regardless of culture.

30. To learn more, see Steven Laureys and Giulio Tononi, *The Neurology of Consciousness* (London: Academic Press, 2008); Jean-Pierre Changeux, *The Good, the True, and the Beautiful* (New Haven, Conn.: Yale University Press, 2012).

31. A. Berthoz and I. Viaud-Delmon, "Multisensory Integration in Spatial Orientation," *Current Opinion in Neurobiology* 9 (1999): 708–712.

32. The vestibular system is the principal sensory system for perception of movement and orientation of the body in space. The sensory organ is in the inner ear, which is the source of the sense of balance.

33. This is the reason why patients suffering from vestibular pathology are sometimes regarded as strange, for their description of their symptoms is difficult to understand. Their giddiness and imbalance must be seen just as much as disturbances of perception of space, with illusions of movement or rotation, as of the spatial unity of body and mind.

34. As proof of this opening the Dalai Lama stated that if one day science discovered something that contradicted religion, then religion would have looked again.

35. These fundamental questions were tackled in a collective work by the Dalai Lama and a group of scientists invited for a week to the Dalai Lama's Indian residence. Jeremy Hayward (philosophy), Robert B. Livingstone and Francisco J. Varela (neuroscience), Eleanor Rosch (cognitive psychology), and Newcomb Greenleaf (artificial intelligence) replied to questions by the Dalai Lama and attempted, with him, to build a bridge beyond scientific postulates and religious dogma. J. W. Hayward and F. J. Varela, *Gentle Bridges: Conversations with the Dalai Lama on the Sciences of the Mind* (Boston: Shambhala, 1992).

36. A. Lutz et al., "Long-Term Meditators Self-Induce High-Amplitude Gamma Synchrony During Mental Practice," *Proceedings of the National Academy of Sciences* 101 (2004): 16369–16373.

37. The electroencephalogram (EEG) was developed by Hans Berger (1873 to 1941) between 1924 and 1929. It is a noninvasive graphic recording of electrical activity of the brain, using electrodes on the scalp that measure differences in potential in the circuits of the cerebral cortex. It is the sum of various oscillatory activities characterized by amplitude, frequency, localization, and reactivity.

38. Epicurus defined mind as the cry of the flesh. See also M. Williams et al., *The Mindful Way Through Depression: Freeing Yourself from Chronic Unhappiness* (New York: Guilford, 2007).

39. B. K. Hölzel et al., "Investigation of Mindfulness Meditation Practitioners with Voxel-Based Morphometry," *Social Cognitive and Affective Neuroscience* 3 (2008): 55–61.

40. M. W. Donald, *A Mind So Rare: The Evolution of Human Consciousness* (New York: Norton, 2001).

41. See also M .W. Donald, *Origins of the Modern Mind: Three Stages in the Evolution of Human Cognition* (Cambridge, Mass.: Harvard University Press, 1991).

42. Edgar Morin, *Le paradigme perdu. La nature humaine* (Paris: Seuil, 1973).

5. The Brain Under Repair

1. Montaigne, *Essays*, 1:26.

2. G. Lantéri-Laura, *Le cerveau* (Paris: Seghers, 1987).

3. He was right handed but had to write his memoirs with his left hand after a stroke resulting in hemiplegia. He disagreed with the specialist about the localization: at autopsy he was proved right.

4. J. Olesen et al., "The Economic Cost of Brain Disorders in Europe," *European Journal of Neurology* 19 (2012): 155–162. http://1mind4research.org/news/europes-brain-disorder-bill-hits-800-billion-euros.

5. In 2011, mental illness in France represented a tenth of health costs (11 billion euros), almost as much as cardiovascular disease (12 billion), and was the top expense in hospital care. In Europe, direct costs for mental pathology are higher than for cancer or diabetes.

6. R. D. Terry, R. DeTeresa, and L. A. Hansen, "Neocortical Cell Counts in Normal Human Adult Aging," *Annals of Neurology* 21 (1987): 530–539.

7. D. C. Park and N. Schwarz, *Cognitive Aging: A Primer* (Philadelphia: Psychology Press, 2000).

8. According to the National Institute for Statistics and Economic Studies, in January 2011 the life expectancy for men in France was 78 and for women 85.

9. Neurofibrillary degeneration consists of accumulations of fibrils formed of filaments invading the neuron. These fibrils are formed of microtubular tau protein modified by hyperphosphorylation.

10. It is estimated that about 10 percent of Alzheimer patients suffer from the hereditary form.

11. As there is no reliable biological marker for early diagnosis, when the putative diagnosis is made neurological changes have already been occurring for several years.

12. R. Kawashima et al., "Reading Aloud and Arithmetic Calculation Improve Frontal Function of People with Dementia," *Journals of Gerontology Series A: Biological Sciences and Medical Sciences* 60 (2005): 380–384.

13. Manic-depressive (bipolar) disorder is characterized by periods of mental excitation (*mania* in Greek means madness) alternating with periods of deep depression (melancholia).

14. In 1952, biological psychiatry was born with the discovery of the first neu-roleptic, chlorpromazine (Largactil). Henceforth medication could replace the neurosurgeon's scalpel, so ready to perform lobotomies. This chemotherapeutic arsenal was enriched in 1957 with the first antidepressant, iproniazid (Marsilid), then in 1960 with the first benzodiazepine, chlordiazepoxide (Librium). With progress in neurochemistry in the 1960s emerged the first monoaminergic theo-ries of depression (inadequate serotonin or noradrenaline neurotransmission) and the dopamine theory of schizophrenia (hyperfunctioning mesolimbic dopamine pathway; hypofrontality).

15. C. D. Vargas et al., "Re-emergence of Hand-Muscle Representations in Human Motor Cortex After Hand Allograft," *Proceedings of the National Academy of Sciences* 106 (2009): 7197–7202.

16. Glucocorticoids play an important role in fear, anxiety, and depression. They often exert their effects on behavior by increasing or decreasing the effi-ciency of certain neural pathways. In man the concentration of cortisol, secreted by the adrenal glands, increases with stress.

17. A program of cognitive remediation for patients with schizophrenia or as-sociated disorders such as psychosis and bipolar disorder (RECOS) was developed in the Department of Psychiatry of the Lausanne University Hospital.

18. For a more exhaustive discussion of cognitive remediation, see P. Vianin et al., "Pertinence d'un programme de remédiation cognitive pour patients schizophrènes: l'hypothèse de la plasticité cérébrale," *Médecine et Hygiène* 61 (2003): 1737–1742.

19. W. D. Spaulding et al., "Cognitive Functioning in Schizophrenia: Implica-tions for Psychiatric Rehabilitation," *Schizophrenia Bulletin* 25 (1999): 275–289.

20. M. M. Merzenich et al., "Temporal Processing Deficits of Language Learning–Impaired Children Ameliorated by Training," *Science* 271 (1996): 77–81.

21. P. Tallal et al., "Language Comprehension in Language Learning–Impaired Children Improved with Acoustically Modified Speech," *Science* 271 (1996): 81–84.

22. M. Habib, "Rewiring the Dyslexic Brain," *Trends in Cognitive Sciences* 7 (2003): 330–333.

6. The Enhanced Brain

1. Unless it is the opposite. What if man became so dependent on such technol-ogy that he became enslaved by it?

2. We may recall the maxim "that a robot is not quite a machine. . . . A robot is a machine that is made as much like a human being as it is possible to make it. Arthur Byron Cover, *Prodigy—Isaac Asimov's Robot City*, book 4 (New York: Bryon Preiss Visual Publications, 1988).

3. F. Cabestaing and A. Rakotomanonjy, *Introduction aux BCI*. Actes du 21eme Colloque sur le Traitement du Signal et des Images (GRETSI 07) (Troyes: Uni-versité de Technologie de Troyes [UTT], 2007).

4. We have obtained much of our information from the review F. Lotte, A. Lécuyer, and B. Arnaldi, "Les interfaces cerveau-ordinateur: Utilisation en robotique et avancées récentes," in *Journées nationales de la recherche en robotique* (Obernai, 2007).

5. For the philosopher Henri Bergson, the mind gradually emerges from matter thanks to the discernment required by perception. But this challenges the traditional definition of perception. For Bergson, to perceive is not to judge or know but to be able to act. The act of perception thus places us directly in matter. It is therefore not a reflexive philosophy: one does not leave oneself to go into the world. It is by leaving the indeterminate world that we grasp the world as it is given to us, without doubting it. In opposition to Kant, the subject does not form the world from prior chaos: the world that imposes itself on us is already formed. J. D. Vincent, *The Biology of Emotions* (Oxford: Blackwell, 1990).

6. Robots controlled by thought could enjoy a great success with individuals or enterprises in the twenty-first century. According to the United Nations, in the next few years we shall see robots climbing buildings to clean windows, guarding large cities, making war, or cooking for us.

7. We can distinguish two major families of brain-machine interface. When a patient voluntarily modifies his cerebral activity and the resulting neuronal signals are analyzed it is called *asynchronous*. In a *synchronous* interface it is not spontaneous brain activity that is recorded but rather its response to specific stimulation. The subject receives a stimulus, and the system analyzes his brain's response seen as variation in field potential (the overall electrical activity over a hundred micrometers or so). These evoked potentials control the machine. Because this brain response is an innate characteristic of an individual, a synchronous interface does not need a long period of learning, unlike an asynchronous one.

8. A. Djourno, C. Eyries, and P. Vallancien, "Preliminary Attempts of Electrical Excitation of the Auditory Nerve in Man, by Permanently Inserted Microapparatus," *Bulletin de l'Académie Nationale de Médecine* 141 (1957): 481–483.

9. The retina is the sensory organ for vision. It consists of a matrix of photoreceptors that transform light into electrical impulses for conduction via a complex network of interneurons to the optic nerve and visual centers of the brain. Retinitis pigmentosa affects 100,000 people in the United States, about 400,000 in Europe, and more than 1.5 million throughout the world. This genetic disease progressively attacks retinal cells and leads to total blindness. Macular degeneration, as its name indicates, results from degeneration of the macula, the small central zone in the center of the retina. This causes distortion and progressive loss of central vision. It concerns mainly people aged fifty-five or more.

10. This artificial retina is produced by the American company Second Sight.

11. A. Sterr et al., "Changed Perception in Braille-Readers," *Nature* 391 (1998): 134–135.

12. The title of this section comes from the works of R. Sussan: *Optimiser son cerveau* (Paris: NYP, 2009); *Demain, les mondes virtuels* (Paris: NYP, 2009).

13. F. Cincotti et al., "Noninvasive Brain-Computer Interface System: Towards Its Application as Assistive Technology," *Brain Research Bulletin* 75 (2008): 796–803.

14. A. Lécuyer et al., "Brain-Computer Interfaces, Virtual Reality, and Video-games," *IEEE Computer* 41 (2008): 66–72.

15. For more than fifteen years there has been active research in the United States at the interface between neuroscience and biotechnology. Most laboratories have grants from the Defense Advanced Research Projects Agency (DARPA), attached to the Department of Defense and involved in new technology for military use. To this agency, and Bob Kahn in particular, we owe the first publication, in 1973, on a new protocol to link various networks (such as satellite, radio, and telephones), the famous Transmission Control Protocol/Internet Protocol (TCP/IP) so dear to the Web. More recently DARPA scientists have demonstrated machines (such as robot helicopters and drones) piloted at a distance by brainpower. They have also developed bionic prostheses that considerably improve human performance, such as electronic lenses that enrich vision through information from calculators or detectors. Human beings with such artificial aids using ultramodern armament confer a certain supremacy on the United States.

16. J. D. Bauby, *The Diving Bell and the Butterfly* (New York: Knopf, 1997).

17. P. R. Kennedy and R. A. E. Bakay, "Restoration of Neural Output from a Paralyzed Patient by a Direct Brain Connection," *NeuroReport* 9 (1998): 1707–1711.

18. J. M. Carmena et al., "Learning to Control a Brain-Machine Interface for Reaching and Grasping by Primates," *PLoS Biol* 1 (2003): e42.

19. See also the earlier report: J. K. Chapin et al., "Real-Time Control of a Robot Arm Using Simultaneously Recorded Neurons in the Motor Cortex," *Nature Neuroscience* 2 (1999): 664–670.

20. http://www.ric.org.

21. T. Elbert and B. Rockstroh, "Une empreinte dans le cortex des violonistes: une étonnante plasticité jusqu'aux confins du pathologique," *La Recherche* 289 (1996): 86–89.

22. M. Nicolelis, "La pensée aux commandes," *La Recherche* 410 (2007): 68–72. M. Nicolelis, "Piloter un robot par la pensée," *Les Dossiers de La Recherche* 30 (2008): 30–35.

23. R. Fuentes et al., "Spinal Cord Stimulation Restores Locomotion in Animal Models of Parkinson's Disease," *Science* 32 (2009): 1578–1582.

24. This is the commonest behavioral disorder in children, affecting some 5 to 10 percent. It is generally diagnosed around ages four to six.

25. V. J. Monastra et al., "Electroencephalographic Biofeedback in the Treatment of Attention-Deficit/Hyperactivity Disorder," *Applied Psychophysiology and Biofeedback* 30 (2005): 95–114.

26. These signals are characteristic of the mental state of a subject. For example, at rest with the eyes closed the dominant frequency of brain signals is generally between 8 and 13 hertz (alpha waves).

27. C. S. Stevenson et al., "A Cognitive Remediation Programme for Adults with Attention Deficit Hyperactivity Disorder," *Australian and New Zealand Journal of Psychiatry* 36 (2002): 610–616.

28. J. M. Guilé, "Attention Deficit Hyperactivity Disorder: Can Cognitive Sciences and Psychotherapy Play Together?" *Neuropsychiatrie de l'Enfance et de l'Adolescence* 52 (2004): 510–514.

29. P. N. Friel, "EEG Biofeedback in the Treatment of Attention Deficit Hyperactivity Disorder," *Alternative Medicine Review* 12 (2007): 146–151.

30. S. Dehaene, *La bosse des maths* (Paris: Odile Jacob, 2011).

31. We might also imagine reproduction between cyborgs and their incompatibility to reproduce with humans.

32. Cyborg is derived from "cybernetic organism."

33. K. Warwick, "Cyborg Morals, Cyborg Values, Cyborg Ethics," *Ethics and Information Technology* 5 (2003): 131–137.

34. B. V. Zemelman et al., "Photochemical Gating of Heterologous Ion Channels: Remote Control Over Genetically Designated Populations of Neurons," *Neuron* 33 (2002): 15–22.

35. F. Zhang et al., "Multimodal Fast Optical Interrogation of Neural Circuitry," *Nature* 446 (2007): 633–639.

36. J. Lippincott-Schwartz and G. H. Patterson, "Development and Use of Fluorescent Protein Markers in Living Cells," *Science* 300 (2003): 87–91. L. Groc and D. Choquet, "AMPA and NMDA Glutamate Receptor Trafficking: Multiple Roads for Reaching and Leaving the Synapse," *Cell and Tissue Research* 326 (2006): 423–438.

37. G. Nagel et al., "Channelrhodopsin-1: A Light-Gated Proton Channel in Green Algae," *Science* 296 (2002): 2395–2398. G. Nagel et al., "Channelrhodopsin-2: A Directly Light-Gated Cation-Selective Membrane Channel," *Proceedings of the National Academy of Sciences* 100 (2003): 13940–13945.

38. D. Evanko, "Optical Excitation Yin and Yang," *Nature Methods* 4 (2007): 384.

39. F. Zhang et al., "Circuit-Breakers: Optical Technologies for Probing Neural Signals and Systems," *Nature Reviews Neuroscience* 8 (2007): 577–581.

40. E. S. Boyden et al., "Millisecond-Timescale, Genetically Targeted Optical Control of Neural Activity," *Nature Neuroscience* 8 (2005): 1263–1268.

41. R. Romo et al., "Sensing Without Touching: Psychophysical Performance Based on Cortical Microstimulation," *Neuron* 26 (2000): 273–278.

42. D. Huber et al., "Sparse Optical Microstimulation in Barrel Cortex Drives Learned Behaviour in Freely Moving Mice," *Nature* 451 (2008): 61–64.

43. In 1897 Ivan Pavlov described the principles of the conditioned reflex, which, unlike an innate reflex, is only acquired after a period of cerebral learning.

44. The threshold for perception described here is much lower than previously estimated from classic experiments on cortical electrical microstimulation.

45. C. Bardy et al., "How, When, and Where New Inhibitory Neurons Release Neurotransmitters in the Adult Olfactory Bulb," *Journal of Neuroscience* 30 (2010): 17023–17034.

46. T. Knöpfel, "Expanding the Toolbox for Remote Control of Neuronal Circuits," *Nature Methods* 5 (2008): 293–295.

47. V. Busskamp et al., "Genetic Reactivation of Cone Photoreceptors Restores Visual Responses in Retinitis Pigmentosa," *Science* 329 (2010): 413–417.

48. H. C. Tsai et al., "Phasic Firing in Dopaminergic Neurons Is Sufficient for Behavioral Conditioning," *Science* 324 (2009): 1080–1084.

49. M. E. Carter et al., "Sleep Homeostasis Modulates Hypocretin-Mediated Sleep-to-Wake Transitions," *Journal of Neuroscience* 29 (2009): 10939–10949.

50. V. Gradinaru et al., "Optical Deconstruction of Parkinsonian Neural Circuitry," *Science* 324 (2009): 354–359.

51. This term encompasses any electronic or cellular device with receptors, connections, and microchips that is associated with a body to repair sensory or motor defects of the nervous system. For example, recent progress in ocular implants (artificial retinas) or brain-machine interfaces (communication by thought) allows a glimpse of innumerable perspectives for correcting handicaps.

52. http://med.stanford.edu/news_releases/2004/april/neuroethics.htm.

53. We call "smart drugs" molecules of low toxicity producing a tonic effect aimed at a supposed improvement of mental function.

54. The "enhanced" brain is different from the "repaired" brain, which is one restored to normal function. Nevertheless, although useful, this distinction remains vague.

55. Norbert Wiener (1894–1964) was the founding father of cybernetics and is still seen as a pillar of modern information and communication technology. In 1948 he published *Cybernetics; Or, Control and Communication in the Animal and the Machine* (New York: John Wiley & Sons), which established the concepts of this new field. This science of analogies between organisms and machines formalized the notion of feedback and has implications for engineering, systems control, informatics, biology, philosophy, and even the organization of society.

56. The locus coeruleus (from the Latin for blue) is a small nucleus in the brainstem. It contains half the brain's neurons that use noradrenaline as a transmitter and sends axons to brain regions that we associate with panic and wakefulness (amygdala, hippocampus, and septum, as well as the cortex, reticular formation, and other parts of the brainstem). Its activation stimulates the prefrontal cortex, which is important for control of attention.

57. http://www.modafinil.info.

58. A. Pollack, "A Biotech Outcast Awakens," *New York Times* (October 20, 2002).

59. "Pill to Boost Brain Power," *BBC News* (November 5, 2002), http://news.bbc.co.uk/2/hi/health/2403613.stm.

60. *Through the Looking-Glass, and What Alice Found There* is a novel written by Lewis Carroll in 1871 as a sequel to *Alice's Adventures in Wonderland*. Playing in her house Alice wonders what is on the other side of the mirror. In this new world everything is reversed. She becomes the subject of a chess game involving a Red Queen. During the game Alice and the Red Queen are running at full speed. But Alice notices they are not getting anywhere. "'Well, in our country,' said Alice, still panting a little, 'you'd generally get to somewhere else—if you run very fast for a long time, as we've been doing.'" The Queen replies: "Now, here, you see, it takes all the running you can do, to keep in the same place."

6. THE ENHANCED BRAIN

61. Glutamate is the transmitter released at excitatory synapses. It binds to several subtypes of receptor of which two are particularly important: AMPA and NMDA. The AMPA receptor is coupled to an ion channel through which sodium enters the neuron when glutamate binds, which depolarizes the neuron. If this depolarization reaches a certain threshold, an action potential fires, and the impulse travels to the next neuron. The NMDA receptor is also a glutamate receptor coupled to an ion channel, but it lets calcium into the cell.

62. According to the French Observatory for Drugs and Dependence (OFDT), 4 percent of the French from twenty-four to forty-four years old are consumers of cocaine. 1.5 percent of the French from fifteen to seventy-five consume amphetamines (OFDT).

63. The term "dependence" is used for compulsive behavior in an individual searching for a drug. It is ambiguous, for it can be confused with physical dependence, a sensation of withdrawal in the absence of a substance, as during weaning from a drug. There is therefore confusion between dependence and addiction, although the terms are not interchangeable. For example, cocaine is an addictive substance that does not cause physical dependence. On the contrary, withdrawal symptoms are associated with the use of antidepressants although the patient shows no compulsive behavior toward the drug.

64. The French neurobiologist Henri Laborit discovered the first neuroleptic, chlorpromazine, a relaxant. Later it enabled schizophrenia to be treated without resort to the then popular lobotomy.

65. TREND reports began in 1999 in seven French towns. The 2010 report by A. Cadet-Taïou et al. is a synthesis of observations in 2007 and 2008 and preliminary results from 2009. There are two sections: a part concerning different user groups and contexts and methods of use and a part centered mainly on the broad families of psychotropic drugs (opiates, stimulants, hallucinogens, and illicit use of medication). Besides some fundamental tendencies noted over several years, such as the spread of cocaine in society or the loss of interest in Ecstasy in tablet form, several new phenomena merit attention: the growing social diversity of users of illicit substances, the risks accepted by an insecure young population, and a new cycle of heroin distribution after years of decline following the availability of substitutes.

66. http://www.assemblee-nationale.fr/rap-oecst/drogues/i364113.asp#P198_17242.

67. U. D. McCann et al., "Positron Emission Tomographic Evidence of Toxic Effect of MDMA ("Ecstasy") on Brain Serotonin Neurons in Human Beings," *Lancet* 352 (1998): 1433–1437.

68. H. Greely et al., "Towards Responsible Use of Cognitive-Enhancing Drugs by the Healthy," *Nature* 456 (2008): 702–705.

69. In *Physics* II 1, Aristotle said: "Of things that exist, some exist by nature, some from other causes. By nature the animals and their parts exist, and the plants and the simple bodies (earth, fire, air, water)—for we say that these and the like exist 'by nature.' All the things mentioned present a feature in which they differ from things which are not constituted by nature. Each of them has within itself a principle of motion and of stationariness (in respect of place, or of growth and

decrease, or by way of alteration). On the other hand, a bed and a coat and anything else of that sort, qua receiving these designations i.e. in so far as they are products of art—have no innate impulse to change. But in so far as they happen to be composed of stone or of earth or of a mixture of the two, they do have such an impulse, and just to that extent which seems to indicate that nature is a source or cause of being moved and of being at rest in that to which it belongs primarily, in virtue of itself and not in virtue of a concomitant attribute." http://ebooks.adelaide
.edu.au/a/aristotle/physics/book2.html.

70. G. F. Koob and M. Le Moal, "Addiction and the Brain Antireward System," *Annual Reviews of Psychology* 59 (2008): 29–53.

71. J. Bentwich et al., "Beneficial Effect of Repetitive Transcranial Magnetic Stimulation Combined with Cognitive Training for the Treatment of Alzheimer's Disease: A Proof of Concept Study," *Journal of Neural Transmission* 118 (2011): 463–471.

72. "Nano" is from the Greek for dwarf. In the International System of Units (SI) it represents a unit a billion times smaller than the basic unit (10–9). See http://cordis.europa.eu/nanotechnology/.

73. G. Binnig, N. Garcia, and H. Rohrer, "Conductivity Sensitivity of Inelastic Scanning Tunneling Microscopy," *Physical Review. B, Condensed Matter* 32 (1985): 1336–1338. Gerd Binnig and Heinrich Rohrer shared the Nobel Prize in Physics in 1986 for this work.

74. The scanning tunneling microscope uses a very fine conducting stylus so close to the surface of a sample that electrons can pass from one to the other by a tunnel effect. When tension is applied between the surface and the stylus, an electron current can be detected. Thanks to this current the distance between surface and stylus can be controlled precisely. The amazing resolution obtained allows a view of the arrangement of atoms on the surface. Use of atomic force microscopes allows this analytic stage to be surpassed and offers the possibility of manipulating individual atoms.

75. Chemistry and nanotechnology share many properties. Chemistry analyzes composite substances to reduce them to simple elements or, on the contrary, to synthesize substances from basic elements. So, chemistry organizes atoms to form molecules. Nanotechnology shares that objective, but, while chemistry works statistically on millions or billions of atoms, nanotechnology aims to manipulate isolated atoms to construct objects with hitherto unequaled precision. Nanotechnology respects the laws of chemistry that define the conditions under which atoms can combine or not.

76. "At the atomic level, we have new kinds of forces and new kinds of possibilities, new kinds of effects. The problems of manufacture and reproduction of materials will be quite different." Richard Feynman, in his lecture to the American Physical Society on December 29, 1959.

77. Nano-objects are particles with three nanometric dimensions. If only two dimensions are nanometric we speak of nanofibers (solid) or nanotubes (hollow). The best-known nanotubes are carbon nanotubes, but there also exist boron nitride and natural nanotubes. If only one dimension is nanometric we speak of a nanoleaf.

78. This view corresponds more or less to the definition of nanotechnology by the National Nanotechnology Initiative (NNI) created by the Clinton Government in 2000: "The study of structures, dynamics and properties of systems in which one or more of the spatial dimensions is nanoscopic (1–100 nm), thus resulting in dynamics and properties that are distinctly different (often in extraordinary and unexpected ways that can be favorably exploited) from both small-molecules systems and systems macroscopic in all dimensions." The vision of the NNI is of a future in which the capacity to understand and control matter nanoscopically leads to technological and industrial revolution to the profit of all society. There are four objectives. The first is to develop a research program in nanotechnology to give the United States leadership in the field. The aim is to go beyond interdisciplinary borders to conquer new realms of knowledge. This leads to the second objective, to support transfer of knowledge and technology to create commercial or societal benefits. The third objective is investment to meet the needs in research infrastructure and highly qualified personnel in order to develop this field. The last objective is to ensure a responsible and reasonable development of nanotechnology. Environmental, health, and security risks must be evaluated, and the NNI must communicate with the public and reflect on ethical and legal implications of nanotechnology. http://www.nano.gov/nanotech-101/what/definition.

79. "Transistor" comes from "transfer resistor." Its invention in 1948 opened the way for miniaturization of electronic components.

80. Gordon Moore formulated a law that every two years the density of transistors in an integrated circuit doubles.

81. A great effort will be needed to obtain the public's acceptance of nanomaterials and ensure their successful commercialization. The way forward must involve strict rules, as independent as possible from any notion of profit, accompanied by comprehensive public information about the stakes. To avoid precipitous action over problems of this nature demands reasoned social and political risk evaluation. It is high time to make this understood and launch platforms for information and discussion involving wide public debate. If this chain of information is not established, the danger is a return to systematic mistrust of negative effects of scientific and technical progress. Such a failure would encourage a precautionary principle, a modern form of suspicion toward anything that transgresses "natural" law.

82. B. Aïssa and M. A. El Khakani, "The Channel Length Effect on the Electrical Performance of Suspended-Single-Wall-Carbon-Nanotube-Based Field Effect Transistors," *Nanotechnology* 20 (2009): 175203; Y. F. Ma et al., "Improved Conductivity of Carbon Nanotube Networks by In-situ Polymerization of a Thin Skin of Conducting Polymer," *ACS Nano* 2 (2008): 1197–1204; A. V. Andreev, "Magnetoconductance of Carbon Nanotube p-n Junctions," *Physical Review Letters* 99 (2007): 247204; H. Park, J. J. Zhao, and J. P. Lu, "Effects of Sidewall Functionalization on Conducting Properties of Single-Wall Carbon Nanotubes," *Nano Letters* 6 (2006): 916–919; H. T. Man and A. F. Morpurgo, "Sample-Specific and Ensemble-Averaged Magnetoconductance of Individual Single-Wall Carbon Nanotubes," *Physical Review Letters* 95 (2005): 26801; Z. Yao, C. L. Kane, and C. Dekker, "High-Field

Electrical Transport in Single-Wall Carbon Nanotubes," *Physical Review Letters* 84 (2000): 2941–2944.

83. Organized molecular systems (OMS) stem from auto-organization of molecules inducing long-range order on the supramolecular scale. This organization results from short-range weak interaction far from equilibrium thermodynamic states. The possibility for certain molecules to organize spontaneously when in an appropriate environment opens the way to a new pharmacology. OMS are studied using a combination of soft-matter physics, molecular chemistry, biology, and chemical engineering.

84. B. Tian et al., "Three-Dimensional, Flexible Nanoscale Field-Effect Transistors as Localized Bioprobes," *Science* 329 (2010): 830–834.

85. See the work of Dominique Martinez at LORIA, a computer science laboratory in Nancy: http://www.neuronal-engineering.com/dmartine/.

86. H. Bannai et al., "Imaging the Lateral Diffusion of Membrane Molecules with Quantum Dots," *Nature Protocols* 1 (2006): 2628–2634.

87. See the sites of Antoine Triller (http://www.ibens.ens.fr/?lang=en) and Daniel Choquet (http://www.inb.u-bordeaux2.fr/dev/EN/team.php?team=Dynamics%20of%20synapse%20organization%20and%20function).

88. P. Fromherz, "Neuroelectronic Interfacing: Semiconductor Chips with Ion Channels, Nerve Cells, and Brain," in *Nanoelectronics and Information Technology*, ed. R. Waser (Berlin: Wiley-VCH, 2003), 781–810.

89. P. Fromherz, "Three Levels of Neuroelectronic Interfacing: Silicon Chips with Ion Channels, Nerve Cells, and Brain Tissue," *Annals of the New York Academy of Sciences* 1093 (2006): 143–160.

90. This barrier protects neurons from external influences. It depends on the special structure of cerebral blood capillaries, which act as a selective filter. It is beneficial for the brain by protecting it from pathogens and toxins but it acts as a brake to the action of various drugs that cannot cross it to act on the brain.

91. G. Férone and J. D. Vincent, *Bienvenue en transhumanie* (Paris: Grasset, 2011).

92. Six million people in the world are blind because of infectious disease of the cornea. Unfortunately corneal grafts are still too rare (10,000 per year in Europe) because of the risk of infection from the donor (HIV, hepatitis C, prion disease). So corneal reconstruction from artificial tissue is precious.

93. A bibliography of scientific articles on neurofeedback is available on the site of the International Society for Neurofeedback and Research: http://www.isnr.org.

94. Alpha waves generally indicate times when the subject is calm and relaxed. Some believe they may help bridge the conscious and unconscious.

95. Analysis of the EEG reveals: Delta waves (0.5 to 4 hertz) accompany deep, dreamless sleep. Theta waves (4 to 8 hertz) are manifested when deeply relaxed but awake. Alpha waves (8 to 12 hertz) mark light relaxation and calm wakefulness. Beta waves (13 to 30 hertz) are seen during normal mental activity. Gamma waves (above 30 hertz) mark high mental activity, such as during creative processes or problem solving. The frequency varies according to the type of brain activity. If the EEG is flat it means there is no cerebral activity, one of the definitions of death.

96. Beta waves are dominant when our eyes are open, we listen to music, resolve an analytical problem, make a judgment, take a decision, or process information relative to the world around us.

97. The American Psychiatric Association defines posttraumatic stress as a morbid state of distress following an exceptionally violent event.

98. E. G. Peniston and P. J. Kulkosky, "Alpha-Theta Brainwave Training and Beta-Endorphin Levels in Alcoholics," *Alcoholism: Clinical and Experimental Research* 13 (1989): 271–279; E. G. Peniston and P. J. Kulkosky, "Alpha-Theta Brainwave Neurofeedback for Vietnam Veterans with Combat-Related Posttraumatic Stress Disorder," *Medical Psychotherapy* 4 (1991): 47–60.

99. N. Chevalier et al., *Trouble déficitaire de l'attention avec hyperactivité: soigner, eduquer et surtout valoriser* (Quebec: Presses de l'Université du Québec, 2006).

100. See, in particular, J. Kamiya, "Biofeedback Training in Voluntary Control of EEG Alpha Rhythms," *California Medicine* 115 (1971): 44; J. A. Robbins, *Symphony in the Brain: The Evolution of the New Brain Wave Biofeedback* (New York: Grove, 2008); M. B. Sterman, "Basic Concepts and Clinical Findings in the Treatment of Seizure Disorders with EEG Operant Conditioning," *Clinical Electroencephalography* 31 (2000): 45–55.

101. We might cite the Blue Brain project, a collaboration between IBM and the Swiss Federal Institute of Technology (EPFL) in Lausanne. Using a supercomputer neuroscientists are reconstructing a neocortical column of a rat. Their aim is to simulate the physiology of a human brain and extract the information that makes us unique. Another example is the Human Connectome project, which received forty million dollars in 2010. Among the projects it encompasses is to map the neural networks of the brain by cutting it into very thin sections (one micrometer) with an ultramicrotome, for examination by electron microscopy, followed by computer reconstruction to give a three-dimensional view of the brain.

INDEX

adaptive radiation, 142n7

addiction, 54, 71, 117, 121, 123, 136, 155n18, 165n63

ADHD. *See* attention-deficit hyperactivity disorder

adolescent, 68–71, 94, 120

adrenaline, 95

adrenergic neurons, 39

adult neuroplasticity, 7, 10–11, 104; connectional, 71–73; neurogenesis and, 64–68, 75–77, 156n7, 157n21

affect, 15–16; consciousness and, 58–60; control of, 21

Africa, 2–3, 142n13

aging: of brain, 85–87, 157n23; illness and, 85–86; plasticity and, 70; regeneration and, 63

agnathans, 24, 36, 150nn51–52

allocortex, 43, 152n1

alpha waves, 134–35, 162n26, 168nn94–95

alternative reality, 157n17

Altman, Joseph, 64–65

altriciality, primary, 142n16

altriciality, secondary, 5

Alzheimer disease, 85, 88–89, 97–98, 121–22, 124, 131, 159n10

ampakines, 121

amphetamines, 120–22, 165n62

amphioxus, 24, 35–36, 147nn23–24, 148n31, 150n54

amygdala: role of, 18, 54, 56–58, 71, 95, 164n56; structure of, 15, *15*, 50, 52

amyotrophic lateral sclerosis, 106, 108

analogy, 144n3

ancient brain, 16

animal kingdom, diversity in, 22–27

anticipation, 16, 41

antidepressants, 11, 96, 160n14, 165n63

ApoE4 gene, 89

architect genes, 39

Aristotle, 151n60, 165n69

arthropod, 24, 63, 146n20, 147n27

astrocytes, 75–76, 151n58

asynchronous, 161n7

attention-deficit hyperactivity disorder (ADHD), 109–10, 120

attention processes, 7

Australopithecus, 2